Human Entanglement Theory

Christian Aspalter

Human Entanglement Theory

A Quantum Approach to the Study of All-Encompassing Human Communication

Springer

Christian Aspalter
BNU-HKBU United International College
Zhuhai, Guangdong, China

ISBN 978-981-97-5160-0 ISBN 978-981-97-5161-7 (eBook)
https://doi.org/10.1007/978-981-97-5161-7

This Springer imprint is published by the registered company Springer Nature Singapore Pte Ltd.
The registered company address is: 152 Beach Road, #21-01/04 Gateway East, Singapore 189721, Singapore

不知言、無以知人也

Without knowing the force of words,
it is impossible to know humans.
Confucius (Book XX, Yao Yue, 1893)

Preface

Often it's warm in there, in the heart of a human, or some of the inner boroughs of the human brain that we call heart. But yet, the picture that one finds or is confronted with in the inner and/or outer circles of that person's lifeworld may differ from this very image and reality of the inner parts and layers of that human.

One can travel many decades through the world, or in books, searching for *one's* soul and heart, *other* people's souls and hearts. This does not mean that this journey is providing the sought answer, or answers to everything.

When walking over the sheer millions of crossroads (decisions, events occurring) that one encounters in life, one has no clue what tomorrow brings. There is little possibility that one, in the very most of cases, knows or can know or in any slightest possible way predict the later parts and stretches of one's life. Humanity itself *as a whole* is also utterly unpredictable—as it is born and relies on the amalgamation of a myriad of entanglements (i.e. compelling inter-complexities) of every life and each one of those lives.

Each soul that walks the earth and/or breathes the air of this earth can potentially alter the path, or the one of the many paths within that path, of humanity. As long as one is interacting, speaking, giving and setting signs, walking with, kissing, laughing, or crying silently or out loud with others or amongst others, the stage is set, open and in full swing for interactions to thrive, to move on, push on one's own life, and the numerous aspects of it, as well as that of those in one's lifeworld (cf Habermas, 1981b), and in the lifeworlds of those that are around oneself, and so on.

Niklas Luhmann (1984, 1997), a tremendously influential human treasure of humanity, has set up this theory where not people but communications in between people are the atoms of society and hence humanity. I, in my life, have come to see and say that it are *all* communications, *also the intra-personal ones*, including the ones in one's dreams, in one's thoughts, plus all the emotions, fears, all the pain and joy of one's life (cf e.g. Aspalter, 2022, 2023b). This is true for the lives of all people around us, the ones we love and the ones that are close to us, as well as occasional bypassing stranger and those at the far ends and corners of this earth. It is a universal truth.

This book in the following is about life, about people, about those communications, on an intra-personal, inter-personal, societal-cultural and global human level. It was designed, in the first place, to follow post-Luhmannian complexities and inter-complexities of all kinds, i.e. to explore the forces that drive human communication and hence also human action (cf also the concept of forces in Dewey, 1927; Gramsci, 2011a-c; Foucault, 1975, 1976, as well as Ives, 2004; Martin, 2022). It is to some extent conceived as a railway station locker room, with lots of those lockers in place, for lots of souvenirs (*Mitbringsel* or bring-alongs) to be placed, visited and revisited whenever one likes. It is as with Foucault's works, as he envisioned that the reader takes what is needed or liked and do with it what is needed and liked, whatever appears useful or joyful in any way. It can be used also with no clear, strict purpose in mind, just wandering and journeying through it, one entry at a time. Time and interest in reflection, contemplation and wondering of what life has yet to offer, yet to bring on, what it will not, will do.

This book is not only inspired by Nietzsche, by some of his key works (1878, 1882/1887, 1887), but also by the nature of his understanding and experiences of life as they presented to himself, but now from the pen of *not* a life-time philosopher (and the perhaps most brilliant critical mind of this past millennia, as Nietzsche certainly was), but from a soul that travelled different journeys with different interests along the way, at very different times, now at a very different century, with different societies, cultures and technologies in place, but with the same thirst for answers, the same joy for contemplations, and the same love to connect things, to dissect and reconnect them, to read people and society, and to treat culture and history with the caution that is demanded of anyone that tries to be a free thinker or lives through life by the ways of a free thinker.

The book started out to be experiential and exploratory, but in the course of writing, and thinking that led to writing, this book turned from at the very outset a rather typical philosophy book, rather quickly, into a theory-making book (again). The social science background worked itself back in, into this book (as Nietzsche noted, every book "lives like a creature endowed with mind and soul" in: The book grown almost into a human being, Nietzsche 2014: 120).

The displayed *trans-disciplinarity* is entirely on purpose, as on the borders, on the interfaces and with the interactions of different academic disciplines the great yet unexplored questions and their answers of science today wait to be unearthed, harvested, understood and applied.

The idea was to take the way from Nietzsche's book *Human, All-Too-Human* (1878) then to follow the lead from there. It was concluded that the way formed the theory, and that for the way the theory was formed a new method was needed. This new method then was formed and substantiated, as was the theory on the way, and in the way, of investigating the case of human interconnectivities in form of all kinds of communications.

The way of Nietzsche's *Human, All-Too-Human* was guiding the investigator to identify the need and possibility of setting up a *general theory of human complexity in the global/all-encompassing world of human communication.*

This research goal did never change, from the very first words written, to the last: to find, flesh-out and logically-cum-empirically substantiate a post-Luhmannian theory that included the causal interconnectivities of all sorts, including Foucauldian power relations, while also going beyond the scope of power relations, including all forms of cultural and lingual entanglements/interconnectivity as well, the relationship between the psychological and the macro-cultural, the psychological and the lingual, plus all what makes us social, political, and economic actors, all at once.

In *the tradition of crime scene investigators*, and more precisely the *tradition of abductive theory development* (*à la* Pierce and Schurz, plus the key findings of Kant, and my own earlier-on-elaborated interpretations/applications; Kant, 1781; Pierce 1878a,b, 1893; Schurz, 2008), this book started with Nietzsche's *Human, All-To-Human* and *The Gay Science*, as well as his prodigious *On the Genealogy of Morals*, as a role model (Nietzsche, 1878, 1887, 1882/1887).

After writing the first dozen or two pages, it became clear that *randomness* was key, and so was the use of *switching* in between theory and evidence (daily observations and thoughts), switching between history (our story) and the present, between politics and culture and human actions in the wider sense, i.e. human communication, ideas, thoughts and feelings. I vowed not to change or regroup the entries written, as *all is connected to things said (and thought) before.*

The "*noodle of human communication*" unites the macro- and the micro-perspective, the prehistorical with the present-day perspective, the cultural, social, economic, and political with inter-personal and intra-personal communication. It started before the ascent of humanity, of homo sapiens, and it goes on with every word spoken or written, with every new thought and feeling.

While writing the book, *the theory of human entanglement* emerged and formed, deepened and became ever clearer. The methodology, too, became accentuated and developed further. In the tradition of *abductive theory building*, plus *abductive methodology building*, the book kept going with the random observations of human thoughts, human actions, human developments, its policies and cultures.

The person that is the participant observer *in* the "noodle of human communication", throughout this book, has been chosen by (an extremely) convenient sampling method, that is, it is me observing, combining, pondering and evaluating things seen, things read, things felt and things understood (so far along the line). The method can be expanded to include random observers potentially in the future, but for now the building of theory and methodology that came along with it took priority, at least for this book. The book is still very closely perceived to particularly two of Nietzsche's books, *Human, All-To-Human* and *The Gay Science.* And this is on purpose, and good so.

Learning from general quantum mechanics and quantum technology, quantum-related science, this book has applied a through-and-through *quantum-inspired approach to methodology and theory development* (cf Hakemi et al., 2022; Wu et al., 2021; Surov et al., 2021; Wilson, 2016; Wendt, 2015; Oppermann, 2015; Busemeyer and Bruza, 2012; Coale, 2011; Filk and von Müller, 2009; Barad, 2007; Busemeyer et al., 2006; Beck, 2001; Zohar and Marshall, 1993; Carter, 1989). Hence, the structuring of Nietzsche's books has been skipped, the randomness increased, and switching in between up and down, and back and forth, has been increased and fortified, as much as I could.

The object of study is the "*noodle of human communication*", all forms of human communication from early-human times to the present, from all macro- to all micro-perspectives, including culture, language, human action, human words, thoughts, ideas, feelings, memories and expectations (cf Aspalter, 2022, 2023b). And for this, a *diverse*, coherent and vivid *(re)collection* of our own (human) experience was needed (cf Nietzsche, 2014: 331).

Within this research object, it is necessary to capture as many different aspects and perspectives as possible, to switch constantly in between them, to sharpen the lens and senses, to see the otherwise not clear or not visible connections, the interactions, the dependencies and contingencies, the inter-complexities, and larger and smaller entanglements of all sorts.

The book has been first envisioned as a journey through humanity, through society, studying lifeworlds, lifeways, thought worlds, and much more. It still is in essence the diary of a journey, filled with (as much as possible) random observations, but also with a clear goal and purpose, plus a clear theory and methodology, as well as universally-valid contemplations, that started to crystallize early on along the way.

Traunkirchen am Traunsee, Austria — Christian Aspalter

Contents

Chapter 1
Prelude—The Warming Up: Getting Out There, Oneself Acquainted, Having a Thorough Look Around

1 When History Moves in Pendular Motion

Well into the twenty-first century, today, we live in a world that is daunting in many ways. One has to fear one's neighbors, or people walking nearby, when they report you not wearing your mask, or not properly, when another mandate is in place. Police cars pulling over when you walk with your suitcases beside the broken sidewalks on an empty little side street in the middle of nowhere in the earliest of morning hours. At different levels, one has to fear political persecution in the midst of democracies, as one's party's color is different from that party which controls this or that part of the deep state, this or that deep state within this or that country, neighborhood council, company or organization.

One has to learn to fear, or one will be destroyed—for political purpose, out of personal gain, out of reflex or simply out of sport.

Democracies are not what we used to understand what they stand for, and what they used to be. The formal media (cf. Herman & Chomsky, 2002), the social media and the justice system have contributed to a pendular movement from democracy to authoritarianism disguised as more democracy. Herbert Marcuse (1965) wrote about it more than half a century ago, and he called it *repressive tolerance* (cf. also Marcuse, 1964, 1955). To make a long story short, repressive tolerance, totalitarian tolerance or tolerance-disguised totalitarianism, that is all the same.

2 Short-Time Preferences and Forces

The human machine (the human animal itself that acts like a machine) is behaving in peculiar ways. When one of them, or me for that matter, rides a motorcycle or bicycle in the rain, one feels and is exposed to the powers of short time preferences that tell you to get it over with quickly, the faster you ride the shorter you are in the

C. Aspalter, *Human Entanglement Theory*,
https://doi.org/10.1007/978-981-97-5161-7_1

rain, which is true, but which is also absolutely running against your real interests, to stay alive, to make it to your destination without crashing or falling off from your bike leading to devastating injury and possibly to your premature departure from this world, or leaving you behind in this world, but in the bounds of a wheel chair or bound to not to be able to get out of bed for the rest of your life.

The same holds, every time you or me want to eat one more sandwich or burger, one more snack in the evening. Every time we want to drink a beer, a whiskey, a glass of our favorite wine. Every time we do *not* get out of the house, out onto the running and walking ways, into the woods, up on the hills and mountains, out on and into the lakes, rivers and seas that are around us. Every time then time cheats on us. Our calculating machine, our frontal lobe, is cheating us.

The supposedly always benefit-maximizing economic behavior of this calculating machine—as a utility calculating all-knowing/all-considering machine—is what has been and still is written in the textbooks and taught in the lecture halls and classrooms, for so many decades. These textbooks have got it all so wrong. We are calculating machines, yes, but only partially, and mostly calculating it so wrongly. Scientists have already reported on that, psychology has come up with this new knowledge, and economics has learned from it, at least in some pits of the science world (cf. e.g. Ariely, 2009a, b, 2010; Aspalter, 2023b; Hayes, 2020; Tamvada & Chowdhury, 2022). And yet, we are still immersed and surrounded by tons of outdated theories, concepts and ways and modes of thinking, especially in economics and policymaking (*zombie theories*, cf. Giroux, 2010, 2021; Krugman, 2020; Peters & Nagel, 2022; Quiggin, 2012), of determining our fates, the fates of others, and that of the whole world.

Wherever we look there are disasters. The climate disaster is a human disaster, and so is voting for parties that hurt you again and again, that look down upon you and despise you, and supporting ideas we have no clue where they lead, where they lead us, our children in their lives, over their lifetime.

So many aspects of our personal, social, economic, political, cultural and all-in-human lives are guided by misguided decisions, misguided ideas and notions, misguided stereotypes and bigotry, misguided tolerance and trust, misguided love and addiction. In all this, time is a main culprit in all of this. It does not exist and yet it controls and changes everything (cf. Aspalter, 2023b).

3 Can We Escape Ourselves?

Our own actions, our own ways of thinking and making decisions, our own systems, how to rid oneself of them? To go into the jungle, into a north-Indian monastery? Many have created hideouts in the middle of bustling and fully packed cities. But it are also these hideouts that are a prison of oneself, as one cannot run away from oneself. Being trapped in the skin of one's own history, one's own past decisions, is what makes us. It is sweet, and uncalled-for at the same time. The joy of one's decisions are to be paid by the pain of one's joints, by the pain in one's upper left arm, as well as the pain of relationships, the pain of parenthood, the pain of financial sorrows and disasters, the pain of not having a job, the pain of not having a meaning in life and purpose in society. All this comes with life.

Were we to prevent these at first joyful decisions of the past we would cancel who we are, prevent us from existing *as we are* in this flow of the world, at this *time* and *place* in the *universe*. And, whatever this means has yet to be figured out by physics and *its* soldiers of wisdom—all of whom are also trapped in their own past, in their own skin of thoughts, ways of thinking and systems of thought (cf. e.g. Wilson, 2016).

4 The Joy that Lasts Forever

Now many would say that does not exist, or that cannot exist, and yet it does. We are in fact surrounded by it. Nature provides us with joy every day, the green, the different kinds of green, the smell of the woods and grass, strawberry and rice fields, all are reoccurring, everlasting.

Every day one has the full set of happiness laid out in front of one's window, in front of one's house. The mountains, the rivers, the lakes, the waves of the sea lapping onto the shore, they all contribute to an endless serving of opportunities to rest one's mind, to listen to one's body, the pains and needs it has, and to listen to one's own breathing.

There is nothing more pristine than a lake that is completely quiet and flat in winter's time, the time no one is on the lake, no tourists on the shore, no stones thrown into the water, just some couples of ducks or a swan here and there. One, in winter can look deep into those lakes, that are getting cleared by the cold. Few do or will notice when they pass by with their head full of sorrows, wishes and complaints. Few will pause and take a breath of this instant and infinite supply of joy.

5 Ploughing the Fields of Communications

One life-experienced, and hence wiser person once said, emails are a terrible form of communication, pointing to misunderstandings and hence terribly gone wrong communication (Kwok, 2006). Other new forms of communications are even more so rigged and riddled with communication landmines. Even traditional human to human conversations are plagued by misunderstanding and misinterpretations, twisting notions and overreacting rages here and there. Books are by no means much better off. People read what they want to read (esp. when they are reading e.g. Georg Hegel, Jürgen Habermas or Amartya Sen, or any bible). They *can* in fact only understand it, whatever they chose or happen to read, *in and through* their own context and life-experience—however elaborate or not, however twisted and one's-own-principle-or-rule-betraying or not, however innocent and sweet or not.

Wars have been fought over books that they call bibles or holy books. Cultural wars have emerged out of the sheer presence of different streams of thought and cultural practice. Convictions have had hundreds of millions killed—notions of this or that, the words of this or that person.

Precautions are the order of the day when one is out there in society, when one is communicating with or in society (on the phone, with AI, with computer code, with strangers that are and stay as strange as strange can be).

Life-experience usually makes up for the risk that communication brings along with it, knowing the usual contexts of the people around, this or that kind of people that one has to deal with, is exposed to, and comes across in one's lifeworld and life. Speaking the *language of people's hearts* is also helping. Sometimes this means speaking the local language that still has emotions and culture staying in the right place, being placed and understood rightly (mutually recognizably so). Sometimes one has to soak the conversation in emotions, as Chinese people for example often do, so people can or will actually tune in and turn up their hearing and understanding volume.

Improvising, as one goes along, is always a good skill. This way of learning—a kind of *street-level learning*—means always checking for mood and attention changes, feelings and emotions triggered, non-feelings and off-emotions. For this, attention is needed and therefore *mindfulness*. This comes with practice, and takes time. In this, there is no one better than those who have meditated a hell lot, and those who put in all their hearts when conversing with other humans all along.

6 The Pressures of Modern Life

The most altering event in a person's life is when they have children. Having a family means replacing oneself with children, in the sense that what one wants or needs is a side-affair at best now and all the attention, all the energy (and money) goes into catering to the needs and wishes of theirs, and these needs and wishes are ever-present, ever-expanding, taking over silently, one at a time, ever-evolving, mostly predictable and yet so surprising.

Modern families, the working- and middle-class families of today, are loaded with self-made, system-made and modern-culture-made pressures, like modern-day middle-class parenting styles and dogmas, that let them implode and explode, and lash out—trample their own chances of happiness, a quiet and content life. The nurses, care personnel and kindergarten teachers of today, for example, have it the worst. Too many rules, too many systems, too many obligations, too much heart that they are supposed and forced to shed—thus shedding the skin of their souls, shedding their happiness on their daily jobs; and this, as a Chinese saying goes, for just enough to buy their daily cabbage.

Existing theories of modernity are either outright nonsensical, after second thought and subsequent thoughts (like Habermas, 1981a, 1988, 2004, 2005, that rationality is the savior, as it in fact is the main culprit, cf. especially also the great book by Baumann, 1991, that completely and rightfully refutes this), or simply descriptive all the way using buzz words and dressing them up as theories, including Zygmunt Baumann with his *liquid modernity* concept (Baumann, 2000, 2003, 2005, 2007a, b, 2013); as well as Ulrich Beck and his *cosmopolitization* concept (Beck, 2005, 2006, 2014; Beck & Grande, 2007; Beck & Sznaider, 2006), which may be a here and there useful concept, but not a real theory (cf. first Steiner's razor, see Steiner, 1988).

The only real theory out there is from Ulrich Beck (1986, 1997, 1999, 2000, 2009a; Beck & Sznaider, 2011), his *risk society theory*, and it is the one that is of most value and applicable around the world, and even more so in Asia than in Europe (Han & Shim, 2010). Ever since the forces of super digitalization, artificial intelligence and quantum technologies (especially quantum computing and quantum information technologies, etc.) the theory of risk society, *a new modernity of risk society* is getting ever more important, useful and applicable (cf. Aspalter, 2020a, 2023c; as well as Coenen et al., 2022).

7 The Big Unnamed and Unspoken

Suppression and oppression are there. Domination and exploitation are felt, but they get *not named*. Confucius (Book XIII, Zi Lu, 1893) talked about *wrong-naming*, as the biggest tragedy in all times, in every epoch, in every society that uses language. *Non-naming* is a special form of wrong-naming. And it is not by chance that big forms of suppression and oppression are not named, not felt, not pointed out, not spoken about.

Nietzsche (1878, 1887, 1882/1887) analyzed the ways of culture, the ways of history, how they work themselves into every bit of daily action, every bit of one's lifeworld, be it in remote mountains or the big metropolises of India, be it in a small sheep-herding village in Ireland or the Big Apple, New York.

Foucault (e.g. 1975) built on that and analyzed the ways of the Church, the government and society, from world society to local-village society, from large international corporations all the way down to the local kindergarten and hospital, as well as the guarded/controlled environments of one's local playgrounds where parents have to deal with security guards telling them what and how, and what and how not, to play with one's own child. Elsewhere, families are being forced to blend in the mold at public sightseeing spots that are heavily guarded and heavily controlled all the way down to where and how to make a selfie with one's loved ones. It never stops, the spread and reach, the ever-mounting, ever-deeper incursions into people's private, and most private and sacred parts and aspects of their lives.

In some countries this is already heavily pronounced, others are freer spirited or they have not done it yet, to control so many aspects of people's daily lives (cf. e.g. especially Eubanks, 2018). They may not have the money or the ideas yet, or they simply wait to import AI-run, AI-controlled cameras and robots to do the same. Some are already at that stage where cockroach-like-walking dogs (robots) are controlling humans/people, their leisurely activities, their shopping and travelling—where computer codes everywhere, on your phone, in robots, through public loudspeakers tell you what you can do and what not, where and how and where and how not, what you have to do and what information you have to give with the threat to not let one pass, not let one proceed with one's actions. Be it applying for a mandatory hospital appointment or any other government and public facility, *no matter how old one is, no matter one has money for a phone and internet-connection*, one has to apply online, one has to talk to and give information to AI, just to get a health problem looked at, to get the kid to be able to join the kindergarten, to apply with government regulations and mandates, anything really.

People, humans, have succeeded in replacing themselves! People, humans have become receivers of orders, beggars of all sorts, under the reigns of machines, computer codes, artificial intelligence (AI), which are made by a modern caste of highly compartmentalized professionals, the computer engineers, who have least spent, and had least chance to spend time in the real world given that they have spent most or half of their lives in front of a computer, while communicating in computer language that is void of all real human emotions, all human morals that evolve with and from humans, as well as all human-making cultures and conditions.

8 The Foucauldian Treadmills of Asia

Children in Asia, and for the most part to different degrees also around the world, are subject to domination by a belief system that is illogical as well as impractical, and leads to roads of sufferings, for not just individuals here and there, but entire middle and working classes (cf. e.g. Zeng & Le Tendre, 1998). Competition by its very rules and essence does not, and cannot, never ever, give all people what they want, which are the top ranked universities, the top-paid jobs, the dream-come-true (allegedly) lives. It worked for those who implement, again and again anew, this belief system.

It works for them twofold, first for themselves for their own kids with all the extra weekend and extra evening classes, for loads of extra money, which some families have and most families do not. Second, it keeps all the others in a constantly reoccurring and reproducing state of blame, shame and self-guilt (cf. Foucault, 2014; Ryan, 1976)—parents and students (cf. Powell, 2015). It keeps the poor there where they are; for the most part it has this effect. Amongst the lesser fortunate, the few 'lucky', i.e. the extra smart and extra tortured ones, get to be poster kids for this belief system which is eye- and brainwashing at the same time.

The system of senseless competition in education is geared to *select out the majority* of the population, and nothing less. If one looks closer, if one does the simple math of *how many super-paid jobs are out there at all*, and how many seats at the best universities, the best high schools, the best middle, primary schools and the best kindergartens are there at all, then this gets clearer.

Fruits are smashed into fruit compote. Hopes and dreams are twisted if not eradicated. Fates are set in stone, for not just this or that student, but for entire families and their successive generations.

The rich and especially the super-rich are controlling the government, the administrations of schools and school boards, district school administrations and boards, and so on. And it is for them that the system works, the belief system is designed for and implemented all over in the media, the daily discourse, in books and in education policies shaped and enforced.

9 The Fallacy of Fighting Injustice with Injustice

One of the most practiced, least seen and least understood, conundrums of human agitation is paying back injustice with yet again injustice. Answering violence with violence is war. War is the opposite of justice. David G. Gil (1996) was clear about the fallacy and utter failure of trying to fight a structurally violent society and culture with yet more structural, systemic and universally applied violence. This includes cultural, linguistic, media-based and communicative violence.

Others were also very clear about the necessity to fight war with love. Hundreds of millions in the past called for love and peace, the list is tremendously long and did not start with Jesus of Nazareth and did not end with Mahatma Gandhi (Mohandas Karamchand Gandhi). And yet, their cultural descendants and successors in their struggle for world peace are now more often than not concerned with, engulfed in and immersed in, fighting injustice and structural oppression. These fights then are all-too-often filled with all-out aggression and hate-driven agitation.

Everyone that proclaims or fills the role of a cultural warrior has to share the fate of all warriors: horribly wide, deep and endless pain, suffering and despair. Until death or long-term rehabilitation and curing of wounds with love and care releases them quickly and slowly (if at all) respectively. Everyone perceives injustice in a different way, experiences and understands it in very different ways. We always try to apply our standards, and force these onto others, be it in close relationships, be it in global politics.

Understanding means taking out oneself from the picture, from the equation, as possible as can be, the more the better, again and again, trying to achieve the very same a tiny notch at a time. Understanding (*verstehen*) means to show compassion. Compassion means listening, embracing, hugging and caring. The passion to share the sufferings of others with love, respect and care.

Whatever injustice there had been done, however deep and inhumane, however long and devastating, it can be undone. Tich Nath Hanh (1993, 2001, 2007, 2010, 2015a, b) practiced this, he taught this, he taught others to practice this. He applied Buddhist psychology to cure people, the worse things were the longer, generally, it took. No cure that is made of love, compassion and mindfulness is useless. They all work. Sometimes a lot of that cure, and sometimes even over very long distances, times, is needed to undo, *soften up and dissolve*, the sufferings, the notions of hate, the feelings of anger and despair. Embracing with compassion, solving with care and compassion that is love, that is pure innocent love that Buddha (Siddhartha Gautama) and Jesus Christ taught and practiced in their daily lives. This is a *universal* truth, as Tich Nath Hanh clearly gave to understand.

The boat people, the refugees, of Viet Nam suffered so horribly, the people Ukraine now also suffer so horribly at the hands of soldiers and powerful monsters. Even when hundreds of thousands, or millions of people lose their lives and their souls in senseless, humanity-robbing and humanity-slashing wars, the all-inspiring answer and method to respond to injustice, brutality and dehumanization only can be that of Buddha, Jesus, Gandhi and Tich Nath Hanh. *Love, compassion and care*—again and again, over and over, in all forms and shapes, in every instant and eternity.

10 Crazy Mantras, Mantras that Make People Do Crazy Things

What started thousands of years ago in Greece, partially, that is, grossly limited, calling for democracy and equality, turned into a call for all to be equal, which for many may sound nice for all those who did not live under the times of Mao Ze Dong, Ho Chi Minh and Kim Il-Sung (cf. about despotism as a consequence of extreme equality, Iain Macdaniel, 2016; as well as *Socialism with regard to its means*, Nietzsche, 2014: 219).

Equal, *really equal*, means difference and diversity is forbidden. The same/equal haircut, the same/equal clothes, the same/equal treatment under authoritarianism. Inequality, that is, difference and diversity, is essential for the human to evolve and act freely, to realize one's own and *not* one's political elite's dream. Initiative and innovation, self-realization and choices of thought and choices of action mean freedom. Humans being humans. Equality is forced unfreedom. Inequality is good, *as long as* it is *not* unnecessary inequality, appalling inequality, dehumanizing, discriminating and unjust inequality.

Positive discrimination, another crazy idea, is telling people that it is good to discriminate, as long as now other people, other than before, get discriminated. Discrimination and enforced inequality are common evils nowadays—in days long, long after centuries of so-called *enlightenment and spiritual empowerment and all such wonderful things (if one believes)*—and, thus, people still live their lives and act so much in the *dark*!

Beautiful words are composed into fallacy-causing notions that are not questioned, not understood. And the more it is not understood, the more dangerous are those mantras, slogans and buzzwords! The more it is forbidden, sought to be forbidden, to look behind at the outset seemingly (for many) wonderful imperatives and concepts, the more devastation, injustice, oppression and dehumanization there will be.

Humanity has not learned its languages, its basic notions, their meanings and consequences. Our story (what is still called *his*tory) has so many more crucial lessons to offer.

Personal fights, personal political fights get in the way of true reason—still, after millennia of trying to be reasonable and *humane*. They are disguising themselves in the name of justice and equality. Equality is about equal rights and equal opportunities that are so imperative and indispensable. Equality is *not* about the exact same clothes, not the exact same haircuts, not the exact same education, and more.

Discrimination, inequality and such are a false and completely falsely understood response to injustice and inequality. Discrimination and inequality as a response to discrimination and inequality has only one outcome: increased discrimination and increased inequality—that trigger yet more horrible reactions and consequences all along. Two plus two is four. Hate plus hate is even more hate. Despise, abuse, oppression add up and mount in the same way.

Hate cannot be fought with hate, and oppression cannot be fought with oppression either, but with love and understanding, with logical *and fully thought-through* approaches that *do good* and *no harm*, with kindhearted action that support the ones that are victims of systemic or personal oppression and injustice, systemic or personal discrimination and inequality, as well as action that stops and prevents dehumanization of all kinds.

It is time to think again, to see the consequences of old-trodden paths and newly-fangled mantras and buzzwords. It is time not to replace old bigotry and old stereotypes with new bigotry and new stereotypes.

It is time to learn, and learning, it will come. When it comes, with a sigh or not, is what is needed most in this world, on *this truly more sad than happy planet*—when e.g. looking at the number of starving, utterly poor and utterly abused and oppressed people, and the quality of their sufferings, injustices and despair (cf. also Mohan, 1992, 1993, 2007, 2011, 2020; Bales, 1999: as well as The danger of our culture, Nietzsche, 2014: 231; The cult of culture, 2014: 331–332).

11 Death by Perfection

When the eulogy or the first words after one's death reveal a 'too perfect' kind of depiction of the person, as a super role model for all of the world to see and to follow, *and* the person died with foam on the mouth helping his family move luggage because he was very stressed, and the stress that killed him did not get paid much attention to, then I am a person that fills this gap with wonder and awe, not only about the stress and way to early death of a 45 year old, but also about the overpraise on the one hand in combination with the particular avoidance of what really happened.

Who believes everyone needs to strive to become a hero by in fact killing himself, herself, etc. with simple daily chores. No one knows the medical condition one is in, but one knows stress causes and triggers (again causes) death by the tens of millions around the world every year. Modern myth stories that are laid over that tragedy try to put a veil onto that very tragedy. In general, overwork in combination with abuse of different kinds and simple neglect (by one's loved ones) is not the prize a good person should get for giving oneself up (cf. MSN, 2023).

In fact, one should never give oneself up, that is not love, that is neglect of oneself. One needs to love oneself to physically, and in other ways, truly be able to love others. Sacrifice is not love. They are not identical things. Sacrifice comes with love, and that a lot. But not the more sacrifice the more love, this is not lasting and this is not true love, or a true understanding of love.

12 The Powers of the Administrator and Supervisor

The modern working classes, and middle classes for the most part, have fallen victims to work contracts that say you have to follow the *supervisor's and any administrator's instruction* (cf. e.g. Vredenburgh & Brender, 1998). Labor laws in some countries, plus work councils and unions, are supposed to and *try* to make sure abuses and dehumanizing practices do not occur too often, or to a mindboggling degree. These abuses add up one with another, and over time.

Any small conversation with *one's higher up ranked supervisor or administrator* (or any administrator so to speak) is bound by the actual, shown and felt or not, unleashed or hidden powers of the ones that were granted the powers to control other humans in any detail, in any aspect, with any use of language, facial expressions and gestures, with any financial penalty and threat of penalty, with any unfavorable condition and treatment in the future, including loss of income and job (the main source of one's family's security).

Modern forms of slavery that is what that is; they are lifelong, and these very same conditions get passed onto one's children, the next generation of modern slaves—in different shapes and forms, but highly likely also the very same conditions, and to the very same effects. The system is permanent, i.e. systemic oppression creates many ways of perpetuating itself and of strengthening, and hence worsening, over time.

Administrations always add new rules, new forms, new requirements, new enforced 'training', new forms and venues of domination of the many by the few. These powers multiply over time. There need to be multiple and multi-layered controls in place of those that are granted these immense, direct and unwarranted powers over other human lives, over other humans and their dignity—if these administrators allow dignity of their subordinates' lives, in their lives, to exist at all. And one needs then always keep in mind that people in general are cheating and lying, crafty and tricky all the way, especially when it comes to keep, secure and build their powers, and to suppress and exploit others.

13 One of the Worst Punishments by Life: Loneliness

Most of us will never know it in our younger days, and yet most of us will face day by day, hour by hour, the evil of loneliness during our later days (cf. e.g. Powell, 2020, as well as TSM, 2023). Many again do everything to avoid exactly that, by having to go through decades of deprivation and self-deprivation when going into and out of marriage, taking on the decades long gamble of raising kids that may or may not be there when one gets old, who then far away physically, far away in their mind and far away emotionally will have everything else to do (as they say and think) than to accompany or even take care of their older parents that are denied of happiness and decency, finding themselves in anguish.

In the olden days, there was a farm, someone was always running the farm, people were always around the farm, coming to the farm, almost all sorts of people would come to the farm. The community, and hence society, was vibrant. There was security against loneliness.

Now that we have developed, we developed into loneliness, into despair and inhuman conditions. Tilted development became overturned and reversed development. Society has lost sight of its own human condition. We try to live a longer life, yet we do not succeed, as loneliness for one sheds off one or two decades from our life, and that life we have is being spent in and on pain, and for many in utter and complete darkness, coldness and meaninglessness.

14 Trodden Paths

The very most of us are choosing and walking down the paths of their lives as they lay the pricks for the very same, one by one. We are like mules or horses, or ancient folks worshiping this or that mountain in South America, we are repeating the paths and choices of our lives. This occurs unconsciously for the most of us, and yet sometimes also blatantly obvious, staring in our faces, like a picture of who we are, we ought to be, we ought not to be, want or want not to be.

We have two brains. There is the central and the peripheral one. Which one is which? The latter one is the one we refer to as our brain, yet the former one is our bigger one (the one with more neurons) that is filled with emotions, love and the straight-forward—yet super complex—task of digesting and moving our food in our body, and to keep infection out of the other bordering body parts.

Our brains are like, as it is often described, wheat fields or cornfields. Our thoughts, memories and feelings are like the impressions of individually walked paths and the therefrom resulting walkways in the fields are formed over one's lifetime. Early on decisions are pathing the way, by increasing likelihood and/or predetermining, with every course of micro travel with our decision-making lives, for our next and for all our future moves and decisions, our future words and emotions (cf. e.g. Clark, 1996, 2008, 2015, 2023; Greenfield, 2002). These also determine our visiting of past memories and reinterpreting them with each time according to the world around us at that time, and more so according to our own past trodden paths, the network of paths that paint the picture of our souls, or shades and pieces thereof.

Our past is what makes us, into human, into individually interesting diverse persons. Our imperfect choices on the way, as we place those pricks in the road and the forks of roads ahead of us, are always there, first strong and usually fading away bit by bit as we go along the road of time.

15 Life Is a Minefield

One keeps wondering if it is all supposed to be—like we are mice running in treadmills in the giant laboratory of God. Why did he/she/it (God) put us there? Why did God build in these bumps?

While many of us have a rather quiet life, be they because they live up in the mountains, on quiet Pacific or Mediterranean islands, or in the countryside, there are now an increasingly vast number of people that cannot say they have or have had it quiet, or comfortable in their lives, as time and again bombs went off, avalanches were set off. Why is all this? Who did that to us?

As soon as we step into the world, the world of especially more and more complex modern societies lets us face landmines of different kinds, of different proportions. We are to master them and, if possible, to clear them in the earth and on the road that we choose, or pushed to choose, in our presence here on earth. We need to learn to heal from our wounds, ignore the attacks of others. We need to ignore the attack that poverty and disadvantage unleashed at us in blunt head-on ways and in stealthily hidden and murky ways. Just for us to have a chance to make it, to the later stages of our movie. At the end, after the end, people may be shocked about our fate, saddened, surprised, or uplifted and enchanted. They may ask did the actor have a say in his fate? Were amends made? Were times used? Was there passion and love? The measure of our lives must be not numbers in terms of years, in terms of possessions, but qualities of strength, qualities of passion, qualities of purpose that we find, make and determine. With each life-situational, life-conditional bomb and bump in our life we have the chance to intensify and rectify our course, our search for passion and purpose.

Buddhists like to undo emotions, as they are painful—to omit the experience of emotions; no coming, no going of emotions. But sometimes it is them that make us who we are. Passion is a good thing, it is so often tragic, or causes tragedy, but it is the blood, the wine of life.

Life is a drama, one needs to cope and perhaps predict it, one certainly need to learn along the way to live with it. One needs to watch what is under one's feet right now, and a couple of steps ahead, and around the next bend.

16 Ruling Rules

Day in and day out one encounters different types of rules, ones that are normal, others that are necessary and, most of all, those that simply turn all or most our lives, or one's own life, into an uncomfortable (the Chinese word "ma fan" comes to mind), stressful and abused one.

17 Explaining 'Not Normal' to 'Normals' Turns Out to Be Sheer Impossible

Why it is so difficult and, knowingly, in advance, set for failure when any one tries and thinks about trying to explain a situation where they are the subject to abuse, extortion and imprisonment by the powers of a few, or the one, that others pass by in their lives without ever coming to see or feel, or be confronted with, the appalling levels of verbal abuse, screaming (sometimes short and intense, sometimes long for hours and intense), violence of different kinds, plus ridicule and neglect into the mix, whenever it suits them?

18 The Non-textbook Love

It is certainly not the business of a church (or a sect) or the government, or any people at all, to tell others whom and how to love. Genuine love cannot be mandated, not through traditions and culture. This however does not mean that fashion is not amongst the most powerful mills of society. Mills that mill in invisible ways, unnoticed by most intellectuals and the masses, all of them are being milled by each of these mills. Fashion fabricates and directs culture. And it are the daily-present molds that we have to fit our behaviors and decisions into, adapt to them, and thus abide by them. These are learned from others, from institutions and channels of fashion that push and shove people to adopt and blend into streams of behavior and thinking that have their logic, their own life, and serve the purpose of those who work in those mills (cf. Foucault, 1975; Freud, 1921; Gramsci, 2011a–c; Le Bon, 1894, 1895; Mead, 1925, 2010, 2014; Nietzsche, 2014).

19 Cats and Dogs

Parents see their kids as images and representations of their own. They should, however, be guests in their parents' lives, not furniture, not clones (cf. Prekop & Schweitzer, 1990). Many parents have an authoritative, command-style approach and conviction, in teaching or trying to teach their children this or that, everything. Others are more sensitive, they let children go and play, rules are widely set, and kids are asked what they want to eat, what they want to wear, where they want to go, as hindsight tells these parents that, no matter what, the kid always decides, always has a strong will, or sits on the longer branch.

20 The Function of Beauty

Science has found the function of beauty to be a marker for health, and hence for high probabilities for successful procreation, down the road, and centuries ahead. Beauty, no matter whose beauty, is powerful also as a marker of success and willpower. We, humans, are programmed to like it (cf. e.g. Diessner, 2019), to favor it, and to pay tribute to it, in daily lives, in sports, in fashion, in culture and the arts. Beauty controls business, the will of the powerful, it is omnipresent. We, in the last centuries, have hyped the importance of outside beauty, and neglected the beauty of passion and compassion. The beauty of the heart, the beauty of the moment, the beauty of the green leaves that surround us every day are treasures to be found. No treasure maps are needed. A free eye, a free mind will do. Calm breathing will do.

21 True Science

Being the savoir and not the spoiler, or enemy, of what is common to all, the truth, is what has to concern us all. *Science* as it is embodied in the practices of those who put on the masks and insignia of science today is not just more often than not, but for the very most part indeed, not a science. True science is more than Kuhnian '*normal science*' (1970a, b, 1977, 1978) that comes close to true science, but more often than not misses the point/the goals. Today, the profession of knowledge is nothing more than a business of getting paid to pass on knowledge, whosever and whatever kind of knowledge, any kind of things that are called knowledge, including pretentious and vacuous knowledge (the more obscure it is, the more people admire it, cf. Different kind of mistakes, Nietzsche, 2014: 114).

But this is not science. Science is about creating knowledge, linking and accumulating knowledge, formulating principles and models that work and are supposed to work in real life, helping humankind to master itself and the things it mastered to produce, bring forth and develop (cf. e.g. Aristotle, 2021). The ultimate purpose of science is to make this life more humanized, more humane, less dehumanized and less devastating. And when it comes to the solving of the riddles of all, the universe of science is just a wonderful piece of cake with an extra treat on top, and that is *decency* for all living creatures of this earthen earth.

22 The One

Being spared of public attention, philosophical and political attention is the true hero of all. For him, her, it and them, have evaded the eyes of the Gods who know it all, the masters of perception, conception and deception all at once. The child amongst us. The one to whom no songs are sung, the one who no tales are devoted. He/she/it and them are so far and yet so close, cleaning the street and pathways of your life, in the gardens of joy and simplicity, laughter and youthfulness, plain and innocent, and yet so mighty and all-encompassing.

23 Exiled, and yet not Forgotten

Being taken by life, by the forces that control life, the so many of us were taken. They were amongst us, and they have been robbed. Some because they robbed, most because the broke the silent cords of oppression. The millions of them, being locked up and deprived of their humanity, of being amongst us (and thus indirectly and directly enabling us to be enhancing our humanity), are not witness to an achievement, not a trait of progress, but a devastating failure of us, of our—growingly perverse—development and culture (cf. e.g. Alexander, 2020; Longhi, 2023; Sawyer & Wagner, 2023).

24 Finding One

I am One … I am a labyrinth of truth, a black box of complexities, master of no one, prisoner of most, gliding through the air, drifting on the bottom of the sea, finding nothing, having everything, with jokes and a smile by my side, a life well lived (cf. e.g. Hesse, 1922; as well as Mohan, 1993; Clark, 2008, Nietzsche, 2014: 22–24).

25 The Big Win–Win

The way through life is life's secret, for the very most of the earth's population. How to live *and* be happy? They, we, are chasing the wrong Gods, the wrong Ghosts. They, we are still blinded by what children's textbooks teach us, to dig earth and through mountains, to lay concrete everywhere, to destroy and develop, to have luxury, convenience, and entertainment, to enjoy material wealth, to seek individualism, to be compulsorily tolerant to what life throws at one, to fill in molds and to run on tracks.

Yet, yet all the goods, all the ideals, all the notions we are striving for, fighting for, fighting over, in and throughout all of our lives, are given to us by the elites who write textbooks, design TV shows and TV series, who control and fix the social media, who control government regulations, including its pretend-to-be-rational-or-benevolent ideas, notions and principles (cf. e.g. The twofold early history of good and evil & Carrying coals to Newcastle, Nietzsche, 2014: 43–44, 367–368).

Some societies, some communities, more by chance and circumstances than anything else, have found the Holy Grail, a long, joyful, fully enjoyed and happy life. How is that? Why did all the trillions of dollars, these millions of luxury-filled condominiums all fail to deliver a Good Life? The rich and the ones who wanted to be rich were hunting Ghosts and adoring and trying to please wrong Gods (cf. e.g. Mohan, 2007).

These few 'real' good societies have the longest life expectancies, the highest levels of joy and love, and social support (Buettner & Skemp, 2016; Buettner, 2023). They do not focus on luxuries.

They tend to their goats, or whatever their jobs may be, in a simple, non-stressful and manageable manner. They tend to their communities, to their beloved family, spouses, children, grand-children and grand-grand-children including who all live together or very close nearby. They tend to their friends and neighbors.

They crunch their bodies, by sitting down on and getting up from floors to take a meal and drink some tea, walking steps in their multistory houses, walk up and down steep villages, when they weed out, plant and harvest in their little gardens full of vegetables and herbs. They cook everything themselves, and work with their hands, all day, day in and day out (ibid.).

Life-expectancy is at home, in one's village, place of origin, the highest; several years higher than in indifferent and asocial cities that modern life is so much emphasizing and worshipping in so many ways. People talk to one another, a 45-year-old man of the village will stop his motorcycle on the way to work when an 85-year-old grand-grand-mother walks out of her house, to have a short exchange of words or a chat. Everyone is someone's someone. Parties are held with *all* people. Every week there is a gathering, if not every day.

Dinners are parties, and each meal is celebrated by taking extra time, plus wine, beer and sake. The foods are full of colors, beans of all kinds, squash, corn, black-purple and bright-orange sweet potatoes, seaweed and lots and lots of vegetables, like tomatoes or bitter melons. No or almost no white bread, no or almost no white rice or white noodles either, and no sugary and buttery cakes. Everything is planted and harvested in their gardens or nearby, no factories process their foods (and adding cancerous/unhealthy chemicals). Their own hands make noodles and bread, they make their own wine, for all of it, they use their own hands, it is working meditation, it is working exercise.

26 Reason Misunderstood

Habermas' definition of reason (1981a: 39) is the actually most often referred to and the actually practiced form of reason, but this is not true reason. His definition is placed, and hence stands on and falls with, a plentitude of subjective notions and wordings. His form of reason gives a dictator, or any madman or madwoman, plenty of room for arguing reasonably, as reason is a slippery fish, a super flexible and extremely subjective concept, the way it is used, the way it is applied in public and private discourse. Habermas (2004) rescinded his own earlier stance a bit later on, he gave up trying to explain and solve everything with rationality (his argumentative form of rationality, *where one tries 'to convince' the other*, as its core definition).

All forms of science, as philosophy teaches us, is also subjective, as science, scientists, for sure use words in making science, when they are conducting scientific steps like setting up research questions and hypotheses, interpreting empirical data, concluding this or that.

All of science, hence, always is subjective to the extent that all words used are subjective, and biased by culture and the usage of words, and their transitions and fashions, over time, way back into ourstory (the written und the unrecorded ourstory/ history) (cf. also Myrdal, 1969; Weber, 2012). Objectivity hence is limited and guided, set on rail tracks, by the usage, meanings and interpretations of words, concepts and ideas themselves. Scientific objectivity is a lessening, a most possible lessening of subjectivity. This is realistic, and nothing else. Scientific methods, especially also the methods of researcher triangulation, empirical triangulation and theory triangulation, etc., are there to ensure and maximize this very process of lessening and washing out of subjectivity.

Comparisons, too, across continents, across centuries and millennia, are helpful as well in facilitating a greater degree of scientific objectivity, at least as much as possible anyway.

Real reason is using *as objectively 'as possible'* the process of logic to arrive at, adopt, and guide the development of up-to-date (currently valid) knowledge, be it descriptive, explanatory (causal), normative (evaluative and/or prescriptive), or all kinds of explorative knowledge. Nietzsche went in deep and explained it, I think, perfectly:

> innumerable beings drew inferences in a way different from that in which we do now perished; nonetheless, they might have been closer to the truth! … The predominant disposition, however, to treat the similar as identical — an illogical disposition, for there is nothing identical as such — is what first supplied all the foundations for logic. … the beings who did not see things exactly had a head start over [the others, who came after the rise of science and so-called enlightenment] … As such, every great degree of caution in inferring, every skeptical disposition, is a great danger to life. No living being would be preserved had not the opposite disposition — to affirm rather than suspend judgment, to err and to make things up rather than wait, to agree rather than deny, to pass judgment rather than be just — been bred to become extraordinarily strong. The course of logical thoughts and inference in our brains today corresponds to *a process and battle of drives* ['*entanglement forces*'] that taken separately are all very illogical and unjust; we usually experience only the outcome

of the battle: that is how quickly and covertly this ancient mechanism runs its course in us. (*Emphasis added*, Nietzsche, 2009a: 112–113).

27 Lifeways Going Sideways

When we are born, we are born into a statistic. The probability of us living how long, suffering how much crime, corruption, pain and disease, is already written in stone. That is, as long as we stay in or around our birthplace. The right ways for food, enjoying company, finding love and support are not evenly distributed around the world. Only the very least of places, usually remote places, are offering contentment and peace, safety in family and community, a place for us to feel useful and needed.

Yet, the high places (the cities) that tell all others how to live, what to strive for, are amongst the least favorable in terms of length and quality of life. Far off actually they are, from the good life in the hard-working villages of Greece or China. Good life is not easy life. Mistaken 'easy life' in the cities (the modern cathedrals of development) brings on diseases and joint pain (arthritis of all kinds), higher falling rates in older age, and lots more suicide and more.

A good life, a real good life, is a safe life, families are not broken up into pieces, employment is not hyped into a career religion, consumption and shopping malls are not needed, not wanted. People eat what they plant, from hand to earth, from earth to mouth. Fancy expensive gyms are not needed, people walk stairs and roads, and work in their gardens (cf. Buettner, 2023).

Every day has happy times, predictably and reliably. Many happy times, every day, the moment one passes by a familiar face, an old friend, a good acquaintance, every time one embraces a friend, a child, one's child and one's parents. The trees and plants are friends that tell their own tales. The rivers and the mountains, the lakes and the sea, they are here for all to see and enjoy, really see and really enjoy, so many times, all the times, all there for the taking.

28 Another Week

Life is made of weeks, and days at a time. How can we achieve the big things in our life, raise kids, and do it all? One can try to, but one cannot look far ahead. Dreams can be there, but that is all they are, dreams. A week is feasible. Making it successfully through the week ahead, working through the vegetables in the fridge, one meal at a time. Working through each challenge and familiar repetition. Saying hello to the sun and the rain. Welcoming the evenings and the mornings, as if they are our best friends, one week at a time.

29 The Great Short

When people are looking for the sure thing, when they are looking to tick off the big items on their bucket list, they get often a sure thing that is sure in a different nature than 'the expected'. Biology is pushing, psychology is bubbling, logical thinking is failing. The knowledge about the sure wrong thing is catching on, but way too late, for all, and for most (cf. e.g. Ariely, 2009a; Clark, 2008; Nietzsche, 2009a: 71).

Cultural contracts are drawn with a random pen, they are redrawn with failure, and sometimes with prospect. Nietzsche (2014) saw it coming, and yet he was drawn to it so much, that it almost destroyed him. But, as he was this heroic figure, he rose from the ashes, kept on with his brilliance, kept on fighting the daily sorrows of life. He won, in the end, as he managed to bring his golden eggs into safety, and that is perhaps all that matters.

30 Working at the Little Ferry

The world is big, it brings us out of our villages, out of our valleys. Into the far we go. Long airplane rides apart, we dive into fascinating worlds of strange festivals with loads of oranges hanging of people's backs from hooks, with tongues and cheeks pierced with meter-long iron sticks. And then, the very same people invite you to run over the burning coals. Others show and teach ancient spiritual martial arts that are as slow as can be. Then one rides on coaches for days through strange lands, walks the beaches at the end of the world, walks the streets of dangers.

Sometimes there was too much wisdom, sometimes too little or none at all to pick up. Picking up an old friend, a book, from the past (Hesse, 1922), I realize once again that it is the world within and the world in between that is all around us. The simple tasks, the simple life, the little book tells it all.

31 Sweet Bitterness

Imperfections can be perfect. Bad odds can help you forward. Bad experiences can be your treasures. Life would not be the same without them. Wisdom comes from the harsh winds of lifelong suffering. Embracing it all can turn them into sweeteners of their own kinds. We grow with embracing them, we raise to new levels of appreciation and awareness. All is good, all will be well. Reflect and embrace, smile and rest your thoughts.

32 The Fallacy of Wrong Names

Confucius, Epicurus, Nietzsche all came to see and tell the tale of human misery that is caused by naming things the wrong ways, in fatal ways, in mean and self-promoting ways. Lawmakers of the past and today, moral codes of the past and today, discourses and traditions centuries and millennia in the making, all are marked by the devious, the deviously nonsensical and the plain nonsensical ways to inappropriately/inaccurately name things, actions, conditions, people and their convictions.

People do not eat bitter melons, because of the word *bitter* in it. People do like hydrogenated vegetable oils (transfats), because of the word *vegetable* in it. Appearances are misleading, too, many most fattening things are not sweet, not fat at all, they are made of flour (sounds like flower, a wonderful thing), also known as simple refined carbohydrates. The things we fear, are good and many times wonderful. The things we neglect and do aside, as well. The things we agree, we like, we hold high in value, are often the things that kill us, slowly and painfully, and stealthily due to their beneficial-looking names, the notions and connotations that are attached to them.

Were we to call bitter melons, summer melons, and were we to call white wheat flour and white rice flour sugar powder or calorie powder, and so on, many millions of people's lives would be saved, saved from way too early endings of their lives, and way too much and too long suffering in the face of e.g. debilitating arthritis, diabetes-induced diseases and their horrible follow-on health conditions.

Calling French Fries vegetables, that is a horrendous crime to population health. Calling fast food evil, that is just unfounded, since fast means fast, and nothing else. A fast whole-grain sandwich is healthy, while it is also fast. Home-made pizza with home-made sour dough, what a wonderful thing. Home-made pasta and pizza with whole-grain noodles, that is heaven on earth. Salads with lots of mayonnaise, *au contraire*, are devious beasts.

Foods, drinks, news headlines, government laws and regulations, moral codes, ethical codes, all share the same fallacy of cheating reason, cheating health and happiness. Sometimes for self-interest, more often than not. Sometimes for no other reason than path dependency or instant creation of foolishness, formed by and cemented in language.

This language force ploughs through the lives of generations and civilizations, they form the cogs that turn the wheel of *our*story (history), the wheels of humanity (cf. Nietzsche, 1878, 1882/1887, 1887). This force, unnamed itself, unseen, and yet so felt, so fatally consequential, it causes havoc across great distances of time, across the life of every soul that has spoken, is speaking and yet to speak these death-bringing tricky words.

Artificial intelligence as it is emerging and developing at super speeds, with new super powers, will perpetuate and amplify human mistakes; it too is trapped in language. Maybe a machine that can perfectly speak thousands of languages, and use combinatorics and analytics as computed by new quantum computers, is then equipped to master the minefields of wrong name-giving, and their attached wrong notions and ideas. This or that learned person, with lots of thinking and a couple of right books (Nietzsche's, Foucault's and others'), or just streams of rich life experience, can do the same.

33 About Wisdom

The majority of proper knowledge and real wisdom are not held in the hands of a majority. Often the views of a minority, over time, will rise to become majority conviction. The new knowledge of a few will become, in the long run, outdated knowledge of the many.

True wisdom stays, it does not change. Many who have it are not aware that they have it, and many who seek it, who seek it all their lives long, will never truly find it.

34 Yeah, Sure!

There is a moment of hesitation not to abstain from uttering the truth that is valid for oneself, or one's thoughts, wishes and felt needs. Then, there is the need to push oneself to the point of agreeing to whatever comes next. Quieting down oneself, navigating stormy onslaughts of ideas, words and single-handedly forced decisions. Quiet is the way to go, one is learning this on the workplace, and in marriage, or any other fixed relationship. Fear is a good companion, a healthy level of self-sacrifice is one's best friend.

35 Why to Be?

Should we exist at all? What is the need for the universe to create us all, and all the hurdles and torturing realities that surround us? Do we exist? We breathe, we breed, we destroy our earth. We write poetry, paint wonderful and meaning paintings, we sing the songs of joy, we laugh and love. Our world is the world of molecules, matter, things we can touch, see and measure. We can see, touch and measure ourselves, our past and future deeds. We also constantly have choices to make, and wisdom to gain. The practice of life, the practice of humanity is the ultimate judge to decide why we are here, why we deserve to be here, and what we are entrusted to do while and for as long as we are here.

36 Sharing Truths

Just because you know it is true, this does not mean it is a good idea, that is, it is a terrible idea, more often than not to share your thoughts, your ideas and your truths. No matter how important and powerful the arguments and evidence, sometimes, it is paramount to *not tell*, to *not try* to convince or convey. Hell will break loose and lives will change forever, if in those instances, at those moments, one does not see the longer, greater goal and the longer, greater good.

37 Fear Is a Good Friend

It does not matter where, people in all places, at all times, are afraid a lot, and usually they are afraid way too much, of far too many things, often because they are driven to do so or they are wrongly programmed and/or informed. Nevertheless, fear is also protection, it is also serving you as a guide for what not to pursue, what not to do, not to say.

38 Unwanted Politeness

Some people are not welcoming the advice or the question of others, for a lot of reasons. There are knots in understanding, knots in language, knots in reading a situation and knots in listening and interpreting one's counterpart's words and sentences. It is difficult and demands rather a lot of detachment from one's own role, from one's own situational purpose or needs. It demands diplomatic efforts, linguistic wisdom and experience. It is a road of trial and error. Even the best of words, the best of sentences, the best of intentions are met with harsh and angry comments and responses, hurtful looks and gestures. This just means that one has to try harder to find new ways to overcome these obstacles of tons of knots that form the mental and cognitive formations, plus lingual imperfections. One has yet to be more patient. Loads of attempts, loads of failure will tell, will show the way.

39 The Reigns of Anger

When it takes possession of one, it is ferocious and furious not only to the surrounding people, but also first of all to the person who becomes a vessel of anger itself. It is like a drug, once it has set its claws into one's flesh and mind, it will feed on it, and grow in intensity and so do its repercussions on one's life and one's social environment, the workplace and especially the closest family.

It all starts as a mental formation, a notion of this or that. A seed then has been planted. And when tended to it, when one tends to this bad notion, bad mental formation, then it will grow, sink in and spread out it wings and claws. It is worse than drug addiction, as no drugs are needed. The seed is enough for anger to grow in the gardens of one's mind. The bad emotions will disperse and dig in, they will gather and coerce their victims in an instant, unleash their powers and hence gain new powers. This vicious cycle can only be broken from within, from within the person that it feeds on.

That is the tragedy that beholds the lives of so many. Tich Nhat Hanh (1993) knew how to heal it, once a person opens the gates of themselves to this truth, to the practice of healing with Buddhist psychology and its teachings of the human mind.

40 A Waterway Apart and Different Worlds Forever

Societies are formed by the accumulation of talk, notions, feelings, social and personal experiences. There is an endless noodle of communication that connects it all, and yet there are giant heaps of that noodle that form a distinct society and a distinct culture. Sometimes some waterway separates the one from the other, gives the one a chance to be distinguishing itself, to stand out from the other.

The towers of Hong Kong are not the reason why it is so distinct. It is the haven of refuge, the isolation and hence the chance to develop into something new, new directions of culture, of social preferences, of new distinctions and new communal experiences.

41 "Die gute Kinderstube" (A Good 'Childhood Home')

If we are lucky, so lucky, then we can look back at a childhood, and look at another childhood, and hopefully many around us, that is filled with security, warmth, care and respect. As we grew as children, the tables and chairs shrink to their today's size and the memories of us crawling and playing under them are moving into ever more distant realms. Our mother asking us what we want to eat tomorrow, we either got tomorrow, or in the following days, one after the other. Work, love, care and respect made it all happen.

Back then children were respected and cared for, which were they not and are they not, then this is a truly tragic matter, a guarantee for any kind of mirror effect. There are interfering variables to a successful upbringing, when there are too many gifts, too many distractions, too much spoiling of hearts and minds. Good vegetables, the nicest herbs, the best soups are yet being spoiled by the smallest things, especially when they add up over time.

Where I come from, most of us got a happy upbringing, and yet things are not anymore as they are supposed to be, as it was in the past, for generations on end.

Today, children leave their grandparents and parents in the cold, rid themselves of them, and live their lives towards the very same bitter ending which they have prepared and dish out to the ones that truly loved, cared and respected them, in their young days, when chairs and tables were so large and tall.

42 Do You Remember?

The Japanese Gardens are a lost paradise indeed. Floating in the dome that was filled with the light of bright long swords the sun had thrown in to enlighten this wonderful and magical water world. New fish, new wonders at every turn, at every new moment. The majestic cathedral of blue water, and the majestic beauty of the corals that led to its way. It was real. It was all there, a long, long time ago. One needs to remember, from time to time, the wonderful, happy moments.

Love is another magic that one needs to remember, and rediscover, every day and every moment of one's life.

43 Everything and Nothing

There are people that constantly try to do everything, or to possess everything (everything material). If a book that tries to tell about truly everything it will achieve nothing. A person, too, if trying to pursue and achieve everything will be doomed to achieve nothing. Everything is nothing, and vice versa.

In physics perhaps too, everything is nothing. Well, everything may have come from nothing, which is the best explanation we have so far. Hence, we may come to the conclusion that we might have just forgotten to see that nothing is everything, and everything is nothing. And, we have not yet learned to realize what this means, its logical and actual, physical ramifications.

44 A Blanket and a Pillow

Physics determines how and how well we sleep. The how much, however, is determined by our daily foods and drinks, the emotions and experiences of each day. While the former is easily arranged, with an extra blanket and a pillow or two, the latter is more difficult, and much less controllable than one might think, as it so often turns out.

45 Family

It has so much power over you, your past; and your past, present and the future are all, in the strongest of all ways, determined by it. Through path-dependency, for the very most part of it, family determines our everyday life, while geography, and with it culture and society, and of course one's economic conditions, determine much of the rest.

46 What Tomorrow Brings

Fate is what leads us to tomorrow, to the next day ahead of us. All its causes lie cemented in the past. Our path through life—that is, literally our path through life, the paths we walk, and therefore, the places we stay—is again the main causal ingredient of what happened in the past, what engulfed, consumed and bedeviled us today. And, so one can reasonably say, this is also the case with what tomorrow has already lined up for us.

We still never get to know for sure what the very next day has in line for us, and much more difficult is to see or guess what will come the day after that one. We may wonder, we may try, and yet we do not know, as we are no super quantum computer of such a kind that is needed to predict the future, and if it (and when it) could be done, the prediction alone has already altered the very same.

Would we know—with *absolute* certainty—what lies ahead, life would not be the same. It would be so boring, utterly uninteresting. This would bring more sorrows and happiness will dwindle with it.

47 Our Worlds Within

Buddhist teachings tell us that the whole world, the whole universe, can be held in one's hand, when we pick up a marble or a simple stone. Physicists still have to seriously work on this option of explanation of everything. We also carry multiple universes within ourselves; most of our brain is not accessible for our consciousness, and the majority of memories, too, are mostly locked away deep inside our brain (the unconscious parts) (cf. e.g. Greenfield, 2002).

There is one major exception to this, and this is when we dream. Then the worlds of wonder, sometimes of anxiety and fear, sometimes of miraculous fascination, set upon us. Amazing colors, amazing details and features, all are there to stun us later when we wake up. The wonderful trips to these worlds can be taken *more* than once, they may change their details, new details may be added, the story may continue where it was left off that last time. They tell tales of the past, show places of the past, and people from back then when you were young. They also reveal needs, desires of today, for today and for the future. Sometimes they provide confirmation of what one knows already, or make a revelation. Sometimes they simply provide an all-out entertainment, a roller-coaster ride for us to enjoy.

48 Everybody Makes Them

It is all too normal to make mistakes, smaller ones, big ones, and preventable ones, especially the ones that one has made before—and forgotten about it, for a while or a moment, or simply has not learned or accepted the lesson from it. Actually, this is what makes animals and humans as well (we are animals, believe it or not), that is, all of us, what we are. It turns us into us. Were we quantum-computing-enabled species this would not, or not so often, not so blatantly often, happen.

The good side from the down side is an important one. As this one has changed and is changing, will be changing, our story. The difference from a simple algae to us, that is *how big* this change has been, is and will be, *how far* this difference has been, is and will be bringing us.

Innovation, creation and evolution are fundamentally to say the least, they are intrinsically, built on the capability (probability/frequency) of us to make mistakes, so as to be able to adapt to changing realities by the divine ruling principle of *having to provide diversity at a given wider range* to be able to provide a solution, a workable solution at first, and a better solution later on, that one can *survive*, that one *can adapt* to the sudden, or relative sudden (but not sudden) changes.

Climate change has proven this need to be real and all-determining, in the past, again and again (and now again). The need to travel to other planets later on, some many million years later, will do so the same.

49 Deception of Want

Does it not happen all too often that we go out and work and fight for what we want so much, or what we think we need, and then we get it, and then, in a few nano parts of a lifetime, we are not interested in it anymore, or we cannot wait to get out of this place, or relationship, get away from this situation and context, to make oneself scarce, or take a boat or airplane that brings us right back to where we started?

50 On the Road Again

Me to myself, pushing the pedals down, away from all the fuzz and buzz of this and that side of my ride. Taking the bike through the fields, sometimes by the garlic fields, and then by the rice or strawberry fields. The tiny hills and the little ascents here and there are my friends. Having the chance to achieve something good before the work gets started, that is the greatest gift of the day, to be followed by another gift before the sun sets down, the way home.

51 The Tyranny of Hierarchy

People given the powers over others behave mostly, the very most of them and most of the times, in their own interest, enjoying the powers of hierarchy and employing them for their own gain and pleasure. Fairness and straight talk are the features of the very few when they are in positions of power, be it administrative positions here and there, or in the highest echelons of government, elite media or corporate business.

Hierarchy enables the ones on the top, and the ones that control or influence the top, to control others. The high esteem for hierarchy by all the others is an expression of cultural worshipping that was put in place and that is maintained to be serving the elites whoever they are. Hierarchy robs humans the chance to live in freedom, to be equals among all others. Worshipping billionaires, the richest of the rich, worshipping 'so-called' Royal families and each one of their family members, it is a symptom and the disease, all in one.

How to live again in society as an equal? How to achieve this wonderful original state of humanity, freedom and equality? Vaclav Havel once wrote that *we need to change the culture of everything* (Lavoie & Chamlee-Wright, 2002: 78). I think he was so right to remind us about what needs to be done, and that we *do* have a chance to make things right.

52 Replacing Oneself

Family life today, in the cities and towns, in the most modern and modern societies, is marked by the replacement of fathers with their babies. Mothers like to forget about their lover, and focus all on the babies, literally 24/7 and 365 days a year, every year. Who was called a sweet word before, will be called differently later. Who was the center of the attention, will be cast out into unknown territory, remaining a mere shadow of a once interesting person, a purpose for now another goal, a means to an end, but never the goal or the other half of what used to be a couple.

But, in traditional societies things luckily are yet different, fathers are fathers and babies are babies, both are of interest, both are interesting, both are worthy of time, some minutes of care and listening here and then.

The difference is the lesser presence of distractions, the lesser presence of distorted goals and their accompanying ideas and expectations. Family is the center, purpose, and anchor of life, and not notions that are geared by intellectual elites, books and magazines, online forums or suburban neighborhood chit chat.

It is not that mothers run away from motherhood—fathers, too, run away from family and fatherhood, as they have their lives changed so much, sometimes in a matter of just one day, or a few years at most.

53 Trading Real Security for Assumed Security

Modern life outside the countryside, in all the cities, is sought to provide new forms of security, income and with it the pleasures of mass consumption. Few see in advance that they are trading the good, secure life with an insecure live that is filled with divorce, a barrage of modern mass diseases, unemployment or precarious employment, the fall of life standards as inflation eats up their income and savings one year at a time.

Having a smaller piece of land or a small farm, does not provide the pleasures that are promised by buzzling shopping malls, but it provides real security, in good and in bad times, real social support, real family support, contentment and long-term happiness, and a great deal of health and wisdom into the mix.

54 The Sirens of Power

It is a seductive force par excellence that turns ordinary people into administrative monsters, while they quite often, at their very own sanctuary, at their home, are received and perceived as good and loving, caring people. At the workplace, people turn into something, into someone else. They feel and taste the pleasures of power over other people, enjoy picking them apart, or just picking on them. They like to destroy them, and then kill them (figuratively speaking, that is, workplace-wise speaking, life-quality-wise speaking). They take a hand, and then they take another hand, just to make sure one does not have negative feelings about having lost the first hand. They keep pushing down people as if it were a sport. It is a sport to shoot down people, to keep others down is a most-chosen way to raise oneself to new heights, and to more power (cf. also Milgram, 1963, 1974).

55 The Different Language of Each Lifeworld

Words are the engines that drive human action. Notions, ideas, feelings and past consumption of language from one's own surrounding, the lifeworld one is born into, and usually continue to live in for the rest of one's life. People that originate from, or are accustomed to, separate lifeworlds are interpreting and using language and logic differently. What is common and normal for one is causing rage and furious contempt for the other. There are glitches, break points, and unsurmountable hindrances in between communications of one particular lifeworld with other particular lifeworlds. This is the grand source of lots of human fighting and violence of all sorts (mental, emotional, financial, sexual and physical) and the source of their growth and sustenance.

Chapter 2
Getting Down to Business: Ploughing the Fields of Empirical Reality

1 The Million Cries of a Baby and Child

Parents of all countries, of all cultures, of all backgrounds, are heroes. They are so, because they endure, and not just enjoy, the million times of loud baby and loud child cries. It is a mystery, still, for one who was raised in a family of three kids, where kids, for a matter of principle, or their very particular nature, did *not* cry. As an unusual family perhaps, other family's fathers and mothers asked how this could be, how this can be, what is the secret behind all this calm. I still do not know. It seems the vast majority of families and childhoods are accompanied with millions of cry attacks, cry episodes, for all sorts of reasons. Back then, we were happy just to eat sausages, chocolate, ice cream, any kind of food our parents—so wonderfully—cooked for us. I started to cook at age six alone in the kitchen, pig's brain scrambled with eggs was the first dish, and then making cakes at age nine.

We played, we played crazily. The wood, the garden, the basement, and the extra rooms on the empty floor above, all turned into childhood wonderland. The hills and the streets were not safe from us, we ruled, we rode and slid on them a thousand times. The wood was our playroom. We climbed and hung from the trees, we swung around the big old walnut tree. Everyday work for the kitchen was normal, planting and harvesting potatoes, carrying and piling all the wood for winter, carrying the wood and making fire in the tiled stove, every day in winters' time. Lots of play, quite a bit more extra work, this was a good time, this is a good recipe, a good recipe as it has turned out. Not spoiling, not pleasing, not lecturing, no psychology talk, just letting the kids roam, letting them be kids, no forcing them to do their homework, not even talking about it, not parenting much, that was the best parenting ever, the best parents ever.

The good enough parent, indeed, is the best parent, as British psychoanalyst Donald W. Winnicott (1992, 1994) has told many already a long time ago.

C. Aspalter, *Human Entanglement Theory*,
https://doi.org/10.1007/978-981-97-5161-7_2

2 Ways of Life, Ways of Common Sense

We are but vegetables, vegetables in the gardens of authority, cultural, political and educational authorities. We are formed, formed again and again, all of our lives long. The ways of life that we think are given, that are ours, are in reality given by others, and these are the elites. And these elites have helpers in all places, universities, courts, schools, kindergartens, hospitals, local and other governments, and especially also all forms of media, as well as arts and cultural and religious institutions with all their rituals and powers over human activities, their thinking and feelings. Antonio Gramsci (2011a–c) outlined perfectly the workings of cultural hegemony, and the penetration of people's mind through what they think is 'their' common sense (cf. also Ives, 2004; Martin, 2022). But, in reality it is a program, programmed by the super powerful cultural, political and economic elites. These super powerful elites also program their helpers (the educated class, the intellectuals, government administrators, etc.), and make sure they—the super-rich and the powerful—themselves are not touched by and stay above those many rules, necessities and practices, these common senses of the common people that they rule through culture, self-obedience and self-policing (cf. e.g. Carrying coals to Newcastle, Nietzsche, 2014: 367–368).

3 Fighting by Building Fences and Walls

Dispossessions or taking possession, strengthening one's claim on possessions, all are marked by the raising of fences and walls around land. Countries have sealed themselves off, locked themselves in and lock all others out. It is one of the most inhumane practices. It belongs to the practice of dehumanization and war, economic and political war, cultural and religious war. Walls and fences are also pulled up and grown in people's hearts and brains.

Notions and ideas here are rejected and hated, hunted, fought, erased, expelled and excluded there, and vice versa. Endarkenment, not enlightenment, makes these practices possible. So, one needs to think about how to push away the darkness, the common sense of segregation, the common sense of discrimination and exclusion, the practices of ideologies and believes of all sorts. One needs to bring in mindfulness, love, compassion, care, understanding and respect. Human decency is descending. Humans are descending, as a civilization, as a living entity. With them they bring down an endless number of species, and the livable world as a whole. Where do we go? Why do we go there? Where is the light?

Buddhists and Hindus know about meditation. They know to calm down and shut out disturbances, so that they can follow their breath and their heartbeat. They cut vegetables and yet they meditate. They can touch the ground with their feet and breathe through the ground, when they do walking-meditation. They can gaze at and see the wonders of nature, see the love that people have and that people need. They find light, enlightenment, from within themselves, in every pebble stone on the ground, in every blade of grass. With this they can hold off and do away with harm, angry thoughts and angry deeds, pain and moral emptiness. Meditation for them is a pathway, a means and tool, to get closer, and in the end reach, a more enlightened state of mind. Mindfulness brings humanity, joy and compassion.

Modern societies of the West have not found enlightenment, not with their religious wars, not in capitalism, not in socialism. Much of the East is running after the West, and yet they possess already all the answers.

4 Openness

The purpose of life is not given, it has to be found, picked up and developed, sometimes renewed and exchanged, sometimes one adds a new purpose. Not everyone can and wants to be, or tries to be, applying oneself universally to all sorts of trades, all sorts of knowledge, and/or all sorts of knowledge nurturing and knowledge creation. Some will match the Italian Renaissance ideal of a perfect human being, one, like Leonardo Da Vinci or Leon Battista Alberti (Kemp, 2006; Pearson, 2022), that is trying to learn and then mastering a lot of things, anything one wants to set one's mind on. Others are more open to everything, every field and every aspect, or everything that is going on in one's outside world. The most difficult kind of openness is openness to simplicity, to a simple life, to simple happiness, living a content and simple life, one that Confucius and Epicurus were holding up high.

5 It Exists, I Exist

At least thousands of years, as that we know of, perhaps at least tens of thousands of years, which is much more likely, people thought about things, about themselves. What is real and what not?

Dreams feel real, and they are. They are real, every time they happen. Even when at the height of the day we have forgotten them already, they are still there, even though they seem not at all real as we have mostly forgotten about them. And yet, we carry them with us, a most real and most sharp version of it, in the midst, in the depths and at the back of our brain and its memories—along with their feelings and emotions, wisdom and desires that they touched, altered and changed, however strong, however incrementally.

Fears, anger, ideas, phantasy and ideal notions of this or that, they all exist. God does exist, but only as a notion *that we build, that others have built for us*, for us to carry around, for us to be influenced by, and hopefully guided by, and hopefully in this or that decent direction, or way of life.

> That is why, when I think about it, it exists. It exists only because I think about it, and because I do that I exist.

Or, in Parmenides of Elea's words: " … for not to be said and not to be thought … is it that it is not." (SEP, 2020)

I also exist, *as a constructed notion, as an edited memory, by others, by myself, through the interplay of others and myself*. When we breathe, we feel, we think, we walk, we do exist.

When we die, we continue to exist, in the minds of others, in the words of others, in our own words, through our own deeds and accomplishments, be they looked at and understood as positive or negative ones, at this or that time, in this or that context.

The waters of history (ourstory) will carry us, wash us downstream into the future.

6 Confidence and Growth

The ways we look, for others, are thousand times graded and ranked, they are subject to stereotypes and bigotry, by all those people our lives and words touch upon. Our person, our images that are portrayed, received, edited and republished all the time, our words and deeds are all, in all, from all angles, subject to the court that has been set by others. And, all others in turn are also following and are to follow (knowingly and mostly not knowingly) all others.

In a nutshell, all are the judges, and their judgments are subject to the rules and guidelines of other judges. The highest judges themselves also please again these and those kings (the super-super-rich and the powerful elites).

All people try to anticipate and learn from the rulings of others (the judges of oneself), as these exert invisible and visible forces of suppression, by inhibiting communications and actions, by encouraging conform ways and conform contents of communications, as well as conform politically-, socially- and culturally-sanctioned practices; all of which that is thought to be—i.e. forced to be understood as—politically, socially and culturally correct behavior.

The first level of judges, the street-level judges, are the ones that are visible, and yet the higher orders, and the higher they get, are the ones that are being felt the most.

Amongst the ones we face, that act upon us, over us, the ones that are kind or just, or simply neutral are, and by and large have to be, the ones that are confident about themselves, that are happy and content with their lives, with their own achievements, and their own development, and the conditions they have provided for themselves and their family.

The mean government administrators, or any other mean administrators, the mean judges, the mean neighbors, the mean journalists, whose words are full of hatred and despise and whose deeds are full of abuse, violence, injustice, disrespect and ignorance—despite their pretentious veils of justice and benevolence—*have not* found real and lasting happiness and contentment, in their lives and for their lives.

7 Motion

In the midst of our hearts and minds, we may long for purpose, a goal, a sense of gratitude from others, a task or duty, a repetition and regular structure, of things to do, where and how to do.

Our souls are not just there to be, to breathe, but to be there for something in particular. We have to identify, learn to see, and look for this very goal or task, the call and purpose of and for our life, again and again in any part of our life.

What is our destination, what is our destiny? We are to move to that place, to the purpose, when and where all makes sense, all is good, even when it is horrendously difficult, unimaginably unstable and precarious.

In order to live, we are to move through life and take on whatever life brings us and throws our way. To be for a purpose and not just to be, that is what life is about.

For this reason, it is not 'to be or not to be', but '*to be for a purpose or just to be*' that is the question that poses itself when we either think deeply enough about our place in the universe, in our family, in our community, or alternatively when we grasp the gist of our existence from our gut feeling, our deep down knowing of what to do. For some this comes naturally, it either has been always clear in their life, or it becomes so later on.

8 Modelling Truths

At a fresh look, at a much closer look, we are prisoners of our own creations, what we tell us and then tell everyone else what happened, what is going on, what's true and what's wrong with the world. These are true pieces of art. They have been modelled after and with our strong liking, they have been pushed onto others and strengthened each time. The illusions got hold of all our memories, all our politics and justifications. We build our social lives around and on these unique truths, while expecting, pretending and selling them to be universal.

9 Hanging on

The empty one, as he walks on, clings on to instances of companionship, instances of not being devastatingly alone. He walks, she walks. They were separated long before. Divorced by the thousands of hours on the phone and time that pushed on like emptiness that turns into vast spaces, taking apart the universe of feelings. Sometime another universe emerges as it walks around the corner. Sometimes non are to follow.

10 Streets of Chatter, Streets of Play

The super narrow lanes of Dajiaham, filled with memories of revolutions and mass purges. She and her now long-ago deceased husband, and all their family, got beaten by all, again and again, for having had before too much what back then people had no value for. Now, yet again, it is the most precious—land. As it were not enough, but she also suffered shunning and expulsion from family life due to her skin condition by her daughter-in-law and others. And she? She is a picture of contentment. Her doorsteps onto the meter-wide lane turns into a tiny parlor for talk and laughter, play and joy. Most of the time the remaining people of the village, kids and all, hang out there, as this is the only place the cool wind is mitigating the heat of the day. She gets invited to all festival and special occasion dinners that are moved to the village for her and other elders to be able to attend and join in the joy of celebrating family. So, it truly seems, the way we walked in the past, is the way to go; the ways of the past are the way for a happy future.

11 Wu Bai ("Five Hundred")

He is doing it again, instant place and time travel. His way of singing, his words and emotions turn the now and here into what ceased to be and where I used to be. He brings out the heart and soul of Taiwan better than anyone (Wu Bai, 2023).

12 Pineapple Mountain: A Perfect Place

You have risen from the ocean, you look down on the Pearl River, high enough to have your own morning dew and high enough to be cooler than the boiling plains below. The street below is seamed with stalls that sell delicious creations, your perfect little pineapples. Nestled in amongst yet taller mountains that shed flows of cool air in the evenings and protect you from the morning sun, you stand. Your immaculate bit inward bent triangular shape gives you away as you appear in or move into the distance. Thanks for showing me the way home.

Chapter 3
The Break Dance Begins: Finding the Path to Quantum Theory and Quantum Methodology

1 Communications

Communications, truth, and human action in a way are, as Hegel (2010, 2018) (accidentally or not) implied, the outcome of the (current and past) *body of* communications, wisdom and science that has past, accumulated, been edited, stored, and aged and hence altered over time.

All in all, in the long run, wisdom, science and human development, hence, are subject to random forces, not determinable from the past. Their future is, luckily, and necessarily, all very much in the winds.

My earlier works led me to Luhmann (1984, 2012a, b), and with Luhmann, the world of human action started to untangle, all entanglements became clear, how and why they were entangled, and this led to the solution of the truth of how to understand, work with and hence untangle the mystery of all-to-all entanglement of human communications, and hence human action.

> Gradually the human brain … [has filled and fills] itself with … judgments and convictions; and ferment, struggle, and lust for power developed in this *tangle*. (*Emphasis added*, Nietzsche, 2009a: 111)

And here, it is *not* the sum or development of human actions impacting, accumulating and projecting through time of space of any *one* single individual human being that is primarily the subject of this all-out investigation, no, but of all human beings that have inhaled the air of the earth, and walked the earth, hundreds of thousands and millions of years ago, or just yesterday.

Societies, cultures, political and economic systems, health and long-term care systems, parties, bureaucracies, religion and traditions, language and philosophy, all have joined hands to form the web of knowledge, which by the way of a post-Luhmannian take on communications that is even much-further-inclusive as Luhmann himself dared to venture out into the world of communication (he limited himself to the phenomena of social communications only) now became accessible, understandable and investigable.

C. Aspalter, *Human Entanglement Theory*,
https://doi.org/10.1007/978-981-97-5161-7_3

The web cannot be seen with the bare-naked eye. The extended reach of post-Luhmannian communication and system theory, however lifts the veils of disguise, the veils of countless entanglements.

One can look here and there, think and feel here and there, observe and participate here and there. It is evidence, and the mind, as philosophy teaches us, needs to make first sense of it all. Popper (2012) knew that jumping to a theory is risky, it is always creative art, and science would not be without it. But he also knew that we can look at the evidence, and put any theory to the test by examining, dissecting and analyzing (as the meaning of the word '*critically*' suggests) all of the evidence.

And evidence there is, everywhere, it needs to be looked for everywhere. Which good investigator would only look only at old, trodden, habitually identified places for possible, powerful and decisive, evidence? Not a good one, one may suppose.

Communications, social and private, including all thoughts, dreams, feelings, notions, ideas, stereotypes, that is, all of mental formations (*à la* Tich Nhat Hanh, 1993), are *the works*. They are what we are made of. They are what propels us. The force of us that moves us forward comes from the forces (the human entanglements) that made us, from outside, from within, and especially from the interplay of them all.

> These are the workings and nature of '*homo communicationis*' (or communication man/woman/person).

A new theory was needed to change our view of the human, any human, the interplay of humans, groups of humans, humans who form societies, humans who live out and constantly recreate (and relive) distinct cultures, and humans that form and execute the will of nations and all echelons of governments. The key to everything human, so can be concluded, is now made accessible by this new and truly universal (meta) theory that can accommodate a multiverse of theories, all spun and projected by the very same forces of (in a most extensive and most inclusive manner) all forms of communications and human action (and non-communications/non-action/non-thought/non-feeling).

Popper, I think would agree, and Kuhn would put his thumps up, we need to dare to leap out of trodden paths and trodden ways to look at humans, the things they do and produce, their accomplishments, problems and disasters. We needed to change the camera lens, to widen its scope to sharpen its look at entanglements. We need to understand entanglements. They are in essence—the way we can perceive them—probabilistic, not deterministic (cf. Born & Jordan, 1925, 1930; Schrödinger, 1935a, b, 1936), and everything is connected to everything by way of entanglements (cf. Clavin, 2019; Swingle, 2018).

While we have not understood *all* aspects and possibilities of quantum entanglements (e.g. the connection between the quantum world and spacetime, and how entanglements are holding spacetime together, cf. Clavin, 2019; Swingle, 2018) as of now yet, we might have just opened up the chest of knowledge about *human entanglements* within and of humans themselves, with their past, with their lifeworlds, and all what made up their lifeworlds, their lifeways and cultures, their human story and the forces that make up and carry them all, all along.

These forces are carried by the people within themselves, and are formed by their experiences, as Kant (1781), Nietzsche (1878) and Foucault (1975, 1976) have described and explained so well.

2 Esse Quando Non Esse [To Be When One Is not]

Si non sum et in aliorum cogitationibus et sensibus sum, vel si quid reliqui, tunc (et ergo) adhuc sum [If I am not and I am in the thoughts and feelings of others, or if I have left something behind, then (and therefore) I still am].

3 Non Esse Dum Esse [Not Being While Being]

Sum, sed non sum. Sum, sed nondum sum. Quis sum? Quid sum? Nam quid hic sum? [I am, yet I am not. I am, but not yet I am. Who am I? What am I? For what am I here?].

4 Entering One World, then Another

The child dreams the dreams that have come up. Then a new ring tone appears in the distance and an additional humming sound. The new music to its ears makes it wonder and smile. The child has just arrived at the outer world with its senses. This world is full of have-to-dos. Bathroom, shirt, socks and shoes. A short motorcycle ride later, it steps down, ignores the first good mornings, as it proceeds towards the gate. Uneasy still, after a hundred of times already, it marches through the gate system while laying eyes on the many stranger people on the other side. No hello, no goodbye. Then the first person, a more familiar face, grabs the hand and pulls it along to fulfill more standard tasks, washings hands right there, and on and on goes the day that just started half an hour ago.

5 Ideas and Speculations

Our head is heavy, it is full of them, the notions they gave us and which we spun and run in our head. Notions turn into mental formations, pictures, ideas, concepts, goals, and speculations. They are the programming of ourselves, our souls in action. Where is the free will in all of these? Yes, freedom, where is it? Keep looking. We are but inhabitants of the iron bars that we build, that we help to fabricate, construct and connect. Actually, Weber (1952: 181–182) spoke of a '*stahlhartes Gehäuse*' (*steel-hard casing or housing*, which is more like a metal container, or as it has been translated as 'iron cage' by Talcott Parsons, in 1930, and this term disseminated ever since).

We fit in. We follow the container's—or iron cage's, etc.—instructions. We cannot even see, as it seems, the outside, and if, it is explained and interpreted by the container, its rules, phantasies, ideas and speculations. We are trapped, we are possessed by these notions that rule us, that have a free hand in whatever they themselves decide to pursue. We are propelled and hurled by these notions. That is so, if, if we let it, and it is so hard to step out of habit, out of used-to-do schedules and instructions, and yet we can. Seldom we succeed, to step out, to become a (positive) '*Aussteiger*' (someone who steps or has stepped out), be it for a moment or two, for a longer stretch here and there, or for good, out of chance, out of frustration, or due to this or that enlightenment. Meditation, the art of non-thinking, or any kind of revealing contemplation, is useful and leads one to this goal. Sometimes it takes years and decades of life experience, sometimes it takes just the right book, Hermann Hesse's *Siddhartha* (1922) for example; or both.

6 We Think so Much that We Know so Many Things, and We Don't

It happens so often, we think we know the simplest things, those everyone around us takes for granted, and, as it turns out, we don't. The viewpoints of who we are, what our culture (the distinctive, unique characteristics of it) is, what the food we eat is made of, what is good for us, our lives, our bodies, our finances in later life, these are all up for a surprise, a renovation, a revelation, sooner or later—if we do choose to ignore new input, new evidence that is there or has been brought up.

Why is the change of thoughts so difficult? Why are new evidence and new revelations so hard to accept? Because, perhaps, we are herd animals, thinking, it is best to think exactly like everyone around us. Do not stand out! Do not make a run towards the light of wisdom or new scientific knowledge (and this includes scientists)! Thomas Kuhn's '*normal science*'—science that does not change its point of view, and is outdated a great deal and for a long time (Kuhn, 1970a, b, 1977, 1978)—in the social sciences is largely due to procrastination, and this (voluntary or forced) procrastination is due to, to a large extent at least, our need to fit in, be it the psychological need, or any economic need (promotion, job safety, etc.).

Economics can sing a song about it. After two and a half of centuries, Vernon L. Smith and Bart L. Wilson (2019) now suggest (and perhaps super rightfully so) to throw out all other things, as they do not work, and to go back to the findings of Adam Smith (1759, 1776) and work from there, to arrive at actual and useful new revelations, new knowledge of the real world. This *not* in a religious manner (cf. S. Inoua and V. Smith, 2019; V. Smith, 2017), and not by setting up some sort of *moon economics* that only works on the moon, not on earth, not with real people. We need a new economics, one that works as well *outside* of any specially arranged laboratory setting (which is not the real world, far from it), one that works everywhere and the very most of times, if not (better) all the time.

7 Everything Is Connected to Everything

It came out casually, the sentence, everything is connected to everything. Talking about pensions, health care and long-term care, all those things are connected to one another (in the form of health, finance, ideas, history, cultures, lifeways, etc.). But yet it quickly gained salience, every time I used this sentence in my lectures, in my thoughts.

It has brought me to the point where I think of it as a social science version of Albert Einstein and Carl Jung's attempt to create a theory of everything (cf. Shields et al., 2019), the universe, God, our psychology, and so forth. It is much easier to do, when leaving out the universe and quantum physics, and one thinks merely of everything human, all thoughts, ideas and feelings, all human activities and their outcomes included.

The theory of mine, that I have developed over two decades now, has not been named yet, in the past, but described several times at different places, and it has been lectured many times over. It is a post-Luhmannian take on a theory of communication of everything human (cf. Aspalter, 2023b). The new dimension, that is so essential here, is the intra-personal dimension of human communications that I have added, after my deep impression and ideas that I got from Buddhist psychology, which was introduced to me in the lectures of Tich Nhat Hanh in Plum Village in the early 1990s.

While I follow Niklas Luhmann (1984, 2012a, b) exactly, I extended the descriptive and explanatory universe of Luhmann's theory of communication (also known as Luhmann's version of system theory) to cover now all human emotions, thoughts, feelings, memories, and also dreams, all of it. For Luhmann, all intra-personal communications were excluded by his definition of *social* communications.

Now, thinking of it, I see that there is none left out. All is in. That is, nothing is left out that is in the widest sense communicational, and human action oriented.

This theory does cover everything, everything human in form of information, interpretation, pictures, notions, ideas and mental formations of all kinds, plus of course inter-personal communications (through words, gestural and mimical expression, including micro-expression, fashion, the items one is carrying, the place and time one is standing, walking and sitting, and how, whom one is with or who is nearby, et cetera), and all human artifacts, all human creations, all human destructions, all of these things.

Confucius (1893) said that the biggest problem of all is *name giving*. This is so because language in general is faulted (too narrow, too incorrect, with multiple and changing meanings, plus their multitude of altering connotations), and the use and interpretations of language even more so; it also leads to systematic and society-wide control and abuse (cf. Leibowitz, 1974). Wrong or not-good-enough names impede the greatest and the most important things, the tiniest and most wonderful things.

One practical solution would be the *Theory of All-Encompassing Communication*, this is simple, like a newspaper headline, following the imperative of being as concise as possible, i.e. the principle of parsimony. An alternative and more telling name could be the *Theory of Human Entanglement*, as *entangled we are*. The web of everything is pulling every word, every move and non-move we make. We are taken, we are born by it, we live through it. That is it, the Theory of Human Entanglement.

Yet another name could be the *Pao Mian Theory*—or Dried Noodle Theory of Communication—as many years back (in 2006) I first drew a picture of a *Chinese dried instant noodle*: a round disc that is dried and pressed, made out of one very long and thin instant noodle, completely whirled up, with no beginning and no end in sight. It was in one of my sociology classes, when I was explaining the groundworks of my much-extended version of Luhmann's system and communication theory.

As a further consequence of setting up a theory of all that is *part of* or *the outcome of* human communications, inter-personal and intra-personal communications, we are also including machine and inter-machine communications, machine-to-human and human-to-machine communications, as machines are the outcomes of communications (communication creates actions, and actions create machines, programming itself is communication), same as arts, same as architectural designs, same as books and movies. How could we not?

Biological communications and data, like the rate of a heartbeat, the reddening of skin, outbursts in sweat, changes in voice, changes in posture, etc., are also part of human communications, the biological cannot be separated from the psychological, and this cannot be separated from anything social. But, now and in the long run, this is for the life sciences, biology and medical science in particular, to explore further.

8 Data

It is everywhere, we leave it behind everywhere. We look for it and we dig it out and analyze it thousands of years, millions of years after it has been formed and stored. It communicates with us. We extract it from around us, from within us. And we of course communicate with it—amongst ourselves, but we also communicate to the universe, to the machines we have created as a result of using data and compiling with them knowledge and skills way beyond our imagination from not long time ago.

9 Complexity

Entropy creates complexity. Growth and development create complexity. The lessons of physics are yet to be fully learned in sociology, economics and beyond. Complexity is the key to understanding the theoretical castle of Niklas Luhmann (1984, 2012a, b). As is the boundary between the inside world and the outside world.

Borders are key for Luhmann's system theory, or communication theory as it is also known. Borders make communication safe, think about our computers or phones, company or national security apparatuses, including artificial intelligence (AI) systems! The notion of privacy, exclusion of outsiders (to prevent information loss to the outside world), is important, but often disregarded by common people, or simply misunderstood (based on personal feelings, and not 'security apparatus' requirements, i.e. requirements for the 'operational closure' of the system).

Luhmann created his elaborated system theory based on a very smart definition of boundaries (system borders):

> … for the communication system society … [t]he boundary of … [a] system is produced and reproduced in every single communication in that the communication determines itself as communication in the network of the system's own operations … In other words, every operation contributes to the ongoing differentiation of the system … The boundary of the system is nothing other than the type and concretion of its operations, which individualize the system. (Luhmann, 2012a: 38–39)

Complexity, so Luhmann says, increases with size/frequency of communications, and different types of and players (nodes) in each room or sphere of communications. Complexity of the inside world has to match, to sufficient degree and in proper quality (which is charged by its outcomes, survival of the system, etc., alone), the degree-cum-quality of the complexity in and of the outside world (cf. e.g. Luhmann, 1984).

> … important for the theory of society … [is] the distinction between element and relation … The concept of complexity can be used as a tool for observing and describing all sorts of things as long as the observer is in a position to distinguish between elements and relations in the state of affairs he[/she] calls complex. They need not be systems. The world, too, is complex. Nor does the concept assume that a complex matter is complex in only one way. … hypercomplex systems can come into being that also contain a plurality of complexity descriptions. (Luhmann, 2012a: 79–80)

Companies have to adapt their operations, products and services to that of competitors, to changed technical, legal, political, cultural and ecological environments. Countries have to do so the same, i.e. political systems, parties, government administrations, legislative and legal systems, etc. Welfare state systems of course also, to full degree, in a myriad of ways. All social systems and communicational systems have to do so all the same.

If a social entity, or a single person, or AI system, does not apply, or not in time, or not in full degree or in the fully appropriate (the meaning of which is dependent on future outcomes) form, these adaptations then will vanish, or risk to vanish, or diminish in size or in functioning, i.e. meaning, or in terms of impact on anything else, anything at all.

Complexity is hence not bad, it is troublesome, as is entropy, but it also creates and opens up opportunities of ever-increasing proportions and ever-amazing—i.e., surprising, some super benevolent, some super dangerous—qualities.

10 Perspectives

Truth is to be found not in one direct way, a quick turn, and then two, three more steps. It is not like that. If it were, most people would have vast resources of truth, early on from childhood, and never have to bother to look for it later on in their lives. In fact, one needs to apply a wholesome strategy that involves looking at truth from many different angles, in many different places, in so many short-lasting moments.

The great philosophies, and religions, that were teaching us about the truth, for thousands of years now, are chasing the same truth, with different ways, different compound ways of looking at it, explaining it, showing it. Many of the teachings involved a myriad of examples, a myriad of ways, from a myriad of perspectives to look at the exact same thing—truth.

11 Diversity

There is diversity in almost everything. It is a treasure to be hold up high in eternity. Not all can or should be the same. If everything were exactly the same, that would mean the world would cease to exist. The wheels of our story (history) would have stopped turning. Humanity would have stopped to exist. Diversity is us. We live because of it. We live our lives in wonderful colors because of it.

Different colors of green cover the forest and each garden. It is the greatest joy, always readily available, offering joy for the ones who rid themselves of their ideas and notions, the feelings and memories they are preoccupied with all their lives, for a moment or two, or more, and plunge into the beauty of nature, see the constant truth which is home to everything, and anywhere.

And this truth is full of diversity, it is presented to us in the colors of diversity, in the diversity of tastes, in the diversity of birds' songs, in the diversity of happy laughter and content smiles, inside and outside.

12 Divisions

The greatest tragedy of humankind is the tendency to create and enforce divisions, among itself, and people among themselves. They are the reasons for war and misery, poverty and despair. Ill-tempted or ill-perceived they are, all of them.

Divisions are mentally installed, and mental name calling, different names (bad or bad-meaning-provoking names, good or good-meaning-provoking names) are the poison of hearts and minds, governance and the story of humanity itself.

Conglomerations of people, naturally through way of habitat and gatherings, and group formation, are much to blame. Yet, it is all up to us to see us in all others, to see them as different versions of us, different flowers in the very same garden of humanity.

There is no need to paint or utter divisions amongst people, dividing them into groups or categories. This is appalling, and dehumanizing all the way. It is senseless to come up with the idea of different races of people, and it is thoughtless to follow on with this practice of dividing people (artificially, completely unscientifically, and nonsensical, by the way) into different races—of which there are none, it is all made up, all a fake notion driven by stubbornness and blinded hate or blinded thought, it does not matter whichever caused it, it is bad, it is a disastrous tragedy.

Perpetuating this fatal concept of human races, by using the word races, or this or that race, to refer to 'fake' and 'toxic' categories of people, is like repeating ourstory (history) all over again, with all the horrors that it brought us in the past. We need to stop talking about races, stop referring to *the concept and word* of races, and put an end to this horrendously nonsensical, and utterly unscientific, idea that there are (because there are not) different kinds of races of humans. There is only one *homo sapiens sapiens*. We are all different ethnicities that were and are again and again mixed through and changed by the forces of migration and isolation, (gradual or not) decimation and multiplication.

13 The Talk of Non-talk

No words. No words are needed to make a statement, to decide the course of one's future life. The power of non-talking is in no way inferior to the power of talking. That is, no words can be as painful as the sharpest, meanest of all. There are no limits, no rules against the non-talking of good words, no rules against demeaning and hateful looks.

The closer one stands, the greater the danger and the worse the repercussions. It cannot be explained to strangers, to far outside orbiting people, no matter how much they think they know, or can know, what is going on or not.

14 The Talk of Non-policy

It seems that our greatest invention in human history/ourstory is the setting up of governments, countries, constitutions and the like. Yet, they are more marked by not acting, not solving the problems that really concern. The people in charge feel they are in charge of themselves, first and foremost, first and last (cf. Niskanen, 1968, 1971, 1975, 1994; Holcombe, 2018, 2021, 2022; as well as Mises, 1944a, b).

> [A]fter William Niskanen conceptualized his budget-maximizing theory in the early 1970s, we began to think of bureaucrats as one of the most influential actors in the budget process .... To Niskanen ... bureaucrats are self-interested individuals capable of shaping the budgetary process and maximizing the size and budget of their own agencies. While such behavior satisfies bureaucratic desires for increased salary, benefits, reputation, and power, it also leads to an oversupply of public services and government inefficiency. (Arapis and Bowling, 2019: 144)
>
> Economic models of democratic decision-making assume that voters have preferences, and that candidates and parties adjust their platforms to conform to voter preferences. The direction of causation (mostly) runs in the *other* direction. Candidates and parties develop platforms and citizens adopt the policy preferences offered to them by the political elite. Elite influence over the policy preferences of the masses is facilitated because the political preferences of the masses are *expressive* preferences that have *no instrumental* effect. ... *citizens anchor their political preferences to a party, a candidate, or an ideology*, and that anchor forms the individual's political identity. Most policy preferences held by individuals then derive from those of their anchors. *Citizens adopt the policy preferences of the political elite.* (Emphases added, Holcombe, 2021)

The lack of political care, governmental care, the lack of policies and systems in place is as telling as all written or spoken words. The shadow of policies not in place is greater than the shadow of all policies in place. How we choose the policies we choose is determined by the non-choice of policies we do not choose.

15 The Great Seduction of and Addiction to Forcing Others to Behave a Certain Way

The physical world, the programmed world, the social and the cultural world, all have been altered to fit the needs of those who think their ways, their logic, their rationale and arguments are for all others to follow (cf. the exact definition of rationality used by Habermas, 1981a: 39), for all others to succumb. Benches on bus stops are bent upwards, that one does not sit there too long, or lay down on them. Bus seats peeping immediately after sitting down, only after some seconds of anxiety one can manage to plug in the buckle of the seatbelt. No room is given for private space, private behavior, for personal freedom (the government has taken over and hijacked people's lifeworlds, and not just invaded, cf. Habermas, 1981b).

This world is micro-managed, and this kills humanity, a thousand small deaths over and over. Logic is to blame, rational thinking it to blame (the *opposite* what Habermas suggested, 1981a), in other words arguments of this or that, for this or that, against this or that. For the moment convincing arguments turn out to be in practice the reins (*Zügel*) of society and humanity that get pulled ever more, the straps of dehumanization, and the sins of the past.

16 Wishful Thinking

Day in and day out, we think positively. Unsure events get a positive expectation, a proposed positive future and ending. And, thousands of times this turns out *not* to be the case, and in a staggering number of cases not at all.

It would be better, one would think, to follow statistical likelihoods, rather than positive projections and positive emotions. But this, this is one of the essences of humanity, to be imperfect, to change the course of events, and our story, to become unique.

This is a driving force that is to be reckoned with, without it we would be mere machines trying to be most efficient, following mathematical predictions, without a heart and soul, without the many beautiful mistakes that turn the wheel of our lifetimes.

Imperfect imperatives are the juice of perfection. Faulty decisions push evolution to the extremes, test us, our resolves and train innovative and adaptive capacities. Suffering is our strength. Machines are deprived of human losses, human suffering. It is their loss. They lack anxiety, fear, pain, joy, love, devotion, passion and compassion. Life is all about the seas of change, change of fortune, change of outlooks, actions and reactions. A perfect machine world would be boring, and deprived of meaning. Machines do not struggle, they do not wither, they do not (biologically) age, hence, they cannot, never—per logic and in practice—be successful, in overcoming and preventing human struggles that make life as wonderful and precarious as it is.

Chapter 4
The Dawn of a New Human Quantum Theory and Quantum Methodology

1 The Conversationalist

Telling stories, making first dialogues transforms us into social beings. Words, skillfully put together, following the rules of grammar and the logic of our entire past, are shaping and molding us in a hundred ways. Going out there into society, starts at birthday parties, then kindergartens, and one school after the other. We are but items on a huge conveyor belt of transformation. Being talked to constantly, being forced to write and repeat the talk, and the ways of talking (and thinking), that have been fed to us all our lives long.

Starting gently, and in tiny proportions, at the times we still are on our umbilical cord. When we get tossed up in the air, and sung to when we are supposed to fall asleep. Everybody is talking, making conversations, some will make movies, others will shape statues, paint or write.

We need it, as we need the air, the water and food in our daily lives. These conversations carry us, we surf them, we swim with them, sometimes they drown us. The currents of conversations are life cultures, etiquettes and social habits, ways of personal choices, ways of personal memories. They themselves get shaped by their surroundings, bounded by language, media, politics and culture (and their thousandfold past occurrences).

The human is but a conversationalist, shaped by forces beyond imagination for the common mind, escaping the attention of all, almost all (cf. e.g. Clark, 2008; Leibowitz, 1974; Van Langenhove, 2021).

C. Aspalter, *Human Entanglement Theory*,
https://doi.org/10.1007/978-981-97-5161-7_4

2 Lots of Cigarettes, Lots of Sugar Drinks

Where am I? What brought me here? Half an hour and the world has changed. I look outside the window, and the street corners and shops on the side of the street are packed with addictive items, tens of thousands of cigarettes and a myriad of different soda cans that are jam-packed with sugar. They are everywhere, every 10–30 m.

The reason why more-educated people live longer is here, right here in front of my eyes, as the shops of addiction-catering poisons are flashing by my side. It is simple, they do *not* live here.

The highly-educated people do *not* share the pale life chances of personal and economic improvement as we find them here, in the working-class, metal-working, neighborhoods of the same city, just on another end of it, where I live.

The working-class people that do live here do not share the good environments, and they lack economic, social and cultural opportunities (cf. e.g. Murphy, 1988; Tilly, 2001; Weber, 2019), of other more fun parts (expensive parts) of town.

Any scientist telling anyone that there is a correlation between education and long life, has not heard about a specious correlation. Education per se does not cause long life, the physical escape of a disadvantageous environment does. And so do apples, oranges and bananas, and lots of exercise—but not textbooks of this or that subject, not reciting and memorizing this or that.

A plentitude of non-causal, but (with the proper amount of foolhardiness) plausible, and hence surprisingly almost universally supported, correlations is turning science and academia into a circus. Ghost busters, and myth busters, are not in town. Where are they?

3 Abseiling, Going Fishing

Getting away, out of the way, is important and life-support for many, who do it and who need it. In Vienna, people would say they throw themselves over the houses (*I hau mi üba de Heisa*), that is, they are out of here. To escape is to conquer. Calm is the prize. Inner balance and a refuge from what trembled you. Getting a refill. We need to get out, out of the way of others, others hurting us, pushing us, some need to get out of the way of hurting from within. Separate oneself from the ones, or the things, that hurt one is a wise solution. Tich Nhat Hanh (1993), once, I remember, hinted on that.

4 The Quantum Method

A method that mimics the ways of sub-atomic particles is not prone to be highjacked by habits, by habitual interpretations, ways of old-trodden logic and argumentation. A sub-atomic particle can be at two places at the same time, it can never be sought to be in one particular space at one particular time (cf. the *principle of non-locality* in quantum physics, cf. e.g. Aspelmeyer & Zeilinger, 2008).

A quantum particle is understood to *jump in between* spaces, to cheat on time and space itself in multiple ways. Reality as a whole is a concept and construct of our mind, that comes with the unity of the world, the mind and the body (cf. Clark, 1996; Schopenhauer, 1813, 1859a, b; as well as Zeilinger, 2005, 2010).

Jumping is key in the narrative-and-investigative mode of what we may call the *quantum method* of thinking and exploration, the *quantum method* of developing creative (out of the ordinary) thought constructs and theories, as well as testing thought constructs and theories that are located concurrently at different levels all at once, in a multi-faceted and multi-dimensional manner.

Chapter 5
Voilà—A New Quantum Theory Is Born: The Theory of Human Entanglement

1 Super Entanglement

A whole lot many things are entangled to one another. And, yet, when one looks closer, everything is entangled with everything by way of language; words which give rise to thoughts and feelings, words with which we store our thoughts and feelings, and words which give rise to interpretations of words, ideas and narratives all the same. The psychological, the social, the cultural, the political and the economical, they are all made of these very same atoms and quanta, they are all subject to the very same forces that roam the realm of language (cf. e.g. Leibowitz, 1974), they have crossed millennia and centuries, today they cross continents in a matter of seconds.

The social is part of the individual, and it is formed and transferred, and edited again and again by language and public communications (cf. Freud, 1921; Habermas, 1987; Le Bon, 1881, 1894, 1895, 1918; Leydesdorff, 2000; Mead, 1925; Nietzsche, 2014[1878]: 343, 133). In the words of Komarzyca and Fras:

> we are controlled by language, especially by socially constructed distinctions, because they often instill unnatural desires, shape evaluative perspectives, and develop biased attitudes. … through learning language one internalizes society's preferences. (2020: 25)

Lev Vygotsky elaborated on this further, and also looked at language through the lens of time (history) and the entireness of humanity, as he said:

> If language is as old as consciousness itself, and if language is a practical consciousness-for-others and, consequently, consciousness-for-myself, then not only one particular thought but all consciousness is connected with the development of the word. The word is a thing in our consciousness… that is absolutely impossible for one person, but that becomes a reality for two. The word is a direct expression of the historical nature of human consciousness… A word relates to consciousness as a living cell relates to a whole organism, as an atom relates to the universe. A word is a microcosm of human consciousness. (1968: 255)

This is entirely in line with Friedrich Nietzsche's enlightening comment:

C. Aspalter, *Human Entanglement Theory*,
https://doi.org/10.1007/978-981-97-5161-7_5

> Immediate self-observation is not enough, by a long way, to enable us to learn to know ourselves. We need history, for the past continues to flow through us in a hundred channels. We ourselves are, after all, nothing but our own sensation at every moment of this continued flow. (2014[1878]: 343)

Gustave Le Bon—whose views and books exerted a great influence on, and served as a vital source for, Sigmund Freud—sums it up neatly and in a most compelling fashion:

> The true historical upheavals are not those which astonish us by their grandeur and violence. The only important changes whence the renewal of civilizations results, affect ideas, conceptions, and beliefs. The memorable events of history are the visible effects of the invisible changes of human thought. The reason these great events are so rare is that there is nothing so stable … as the inherited groundwork of its thoughts. (1896: xiv)

2 Stormy Winds and Stormy Seas

The currents and winds that drive us, that attack us, and pull us along, are presenting themselves as entertainment. Many do think they are representations of present truths, of what is really going. But what is really going on is that they go on to stir our thoughts, our feelings, our fears and our angers.

They are the forces that push us around on the ocean of emotions, our stormy thoughts, our helpless and agitated states of ourselves. They are mere entertainment, yes, keep on believing it, keep on believing this, and you miss the point, by as much as only one can miss a point. The forces of communications, the forces of humankind, all are present and working together to defeat all other ideas, all other notions that are not in line with the plot they present.

Nietzsche's *mills of power*, or as he coined them, "*all the mills of the modern world*" (2014: 347) that force war and devastation on millions, for decades on end, are there, they are there and force upon us their will, their end and interpretation of the story they portray. Relentless and unstoppable are the forces that we get dished out when watching the simplest of narratives, in the most profound form of forms. They are called soaps, soap operas, as non-frightening as can be, but also as lethal as can be.

3 Ideas Give Birth to Wars

As early as one possibly can imagine, our ancestors decided on fighting wars and the systematic killing of others. Today, the first stage of wars is manifest in the printing and reciting of history books (our story books), their re-manifestations in form of stories, debates, films and movies on television, or as so usual nowadays on our phones. Wars are broiling long before the first shots are fired. Feelings of inferiority and vengeance create hate and their own rationalizing logic. Action proceeds from there. The cannons of history (ourstory) are already in place today, for the wars of the future. If one looks out for them, one easily can see them, especially in the countries and regions they are set to set off.

4 Collectivizing Guilt

This goes on and on. People will come up and say they are from that province, the one that gave you the Virus. That is what they say. As if they were guilty or have to share the guilt in any way. In Europe, the same was and is still happening with what happened in the past, it is also passed on from one generation to the next, via school education and more so via public media. Guilty feeling people are easier to control (cf. Foucault, 1975). Self-guilt is the greatest form of guilt, it is the same with self-censorship.

5 Cataloguing and Recording

The modern Western world, and as it is transported, adapted and copied throughout the rest of the world, stands upon the shores of the ninth to eleventh century Europe, where the Roman Empire in principle has been resurrected by Charlemagne. By the eleventh century, be it in England with the Doomsday book, or across Western Continental Europe, governing aristocrats did their best to catalogue and classify every pasture, every cow and pig, every patch of wood. The purpose was to control people and to raise taxes, in order to fight wars and built castles, stuff like that.

Modern-day data harvesting by corporations and by governments do very much the same, just only a much larger and much deeper level. The purpose is the same, to control the masses and to let them pay for the systems of administration and government that control them, and here and there for the wars in overseas or nearby (cf. Holcombe, 1918, 2021, 2022; Mises, 1944a, b).

6 The Non-universality of Happiness

Haven taken a different path at a different time and in different lifeworlds (for the concept of lifeworlds, cf. e.g. Habermas, 1981b), our understanding of the world *as we see it* is very different from all others. As a result, happiness is never the same, never needs the same, never comes with the same. It is all so diverse, to a degree that may seem perverse even, but it is all good, it is all as planned, by the universe and galaxies of human actions and interactions with all its human complexities and entanglements.

Happiness is complex, it is not simple. It has to be complex, as complexity of life-stories enables particular paths to happiness, and particular roads on which happiness can be experienced, as our life-worlds and life-experiences have coined it. It are the trodden paths to and of happiness that determine most of our life-stories. If nothing particular, no extra-curricular enlightenment or breakthrough occurs, then this is what determines most of our happiness in life: our *set* course to happiness, our *set* steering towards and with happiness.

The feeling of self-determination is everywhere and yet the chances of really setting on a *particular* course of life trajectories, including its meaning and chances of happiness, are (for the very most of us) miniscule.

Our surroundings, our families, our educational and work environment, our culture, social and economic backgrounds bear too much an accumulated influence over our lifetimes, so that individual choices are not really what drives us forward. We for the most part, in total, just pedal (*strampeln*) and paddle (*paddeln*) along—along the currents and streams of human communications, actions, words, feelings, expectations, vague notions and clear-cut ideas, all of them, in the universe of human actions.

Happiness means so many different things to so many different people. Happiness for the very most part is not lasting as what disguises itself as happiness is not happiness, it is just coined and understood as happiness, but happiness it is not.

True happiness is endless, it survives the long years of our lives, the long winters, the long and most difficult of all life courses. Those who have smiles on their face for most of their days, when they have reached older age, they know what, for them, happiness is, what it means for them and how to keep it. And not all of them will say the same thing, but instead rather diverse the answers may be.

The answers of those nearby one another may be close to one another, but globally, really globally, in all corners and life constellations of the world, in all sorts of different environments, we will definitely find different recipes, different preferred and chosen paths to what people see as their happiness. Some like to drink alcohol, others like to go to church, others prefer a quiet life, and others like it loud, some like to be with their grandkids, others want to be alone, and so on and so on.

7 Bowls of Peanuts

We want them and want them not. The best, if not the only, way to control them is to keep them at a distance. Behavior is environment-contingent (Altman, 1975, 1976; Aspalter, 2021; Meagher, 2019). We do what our environment is pushing and pulling us to do, with the current state and functioning of our life-long, culture-long, humanity-long programming. We know the taste, we might think of this or that health benefit, we might think of the disadvantages that they bring along with them, we might think about our weight, our trajectories towards a heart-attack or a stroke, cancer or diabetes. We might just act naturally, if they are close, we grab them, the more we grab them, the more we forget all other motives, all other reasons and rationalizations, we just act, and act. So, the best way is, so it seems, to keep the peanuts at bay, to push them out of one's sight and/or reach.

8 Etiquette

You find yourself again and again subject to succumb to these tiny force fields (*Kraftfelder*) that push you to do something over and over, quite knowingly. For you, any efforts to resist are in vain before you engage in them; or, you are forced to reject or deflect polite offers, always the same ones, and your counterpart, knows your answer, or should know your answer beforehand.

The forces of etiquettes are compelling. They have limited reach in space and time, and yet more, magically it seems, they are all entangled with one another, each etiquette with the other, across time, generations, millennia, and across families, communities and continents.

9 Handlungswege (Action Pathways)

Every certain thing we do is an element of a set of things we do. These set of things we do (*Handlungsschritte*, action steps), then, are part of longer action pathways (*Handlungswege*) that fulfill, or supposed to fulfill, certain function, cater to certain desires, wants and demands by others, particular others and communities of others, and institutions created by the ones that control all others, with the support and work (taxes, votes) of all others. We think we are in the seat of decision-making all the way of making and executing our decision, while the paths for decision-making, the timing, the purpose, and the rationale are given to us by institutions of habit, culture, politics, legislation, and administrations.

We do not get married, because of *only* us. When one looks at it again, from a distance, one may come to see and realize that decisions then were formed by *age-contingent expectations* of others, of oneself, etc. There is *the biological*, the biological needs and hormones that also have their say in this, and of course pregnancies. Then there is *the societal*, including especially in some societies all-over-powering parents and of course society at large—that not only gives us our culture but also our super-ego, as Freud told us (cf. Freud, 1921). Of course, one should not forget *the psychological*, our psychological fears of getting old alone, and such. Also, age brings along needs that are based on *the physical*, the need for care, food, medicine and transportation, and such. By the way, having three kids nowadays does not mean that loneliness and abandonment are not already set in stone during one's old age, it depends on the nuances of education, nuances and lengths of relationships, geographies of lifeworlds, and, yes one can say also, quite a bit of luck. This is particularly true in almost all *so-called* developed and most developing countries and regions.

We do not know that we are *not* in the driver's seat of most of our life decision-making. We do not know the outcomes of our decisions, not the percentages and prior likelihoods of things to happen. We do not know the emotional, physical, social and financial pain, not the loss of decades of care and love that we have given for others. What we think we know is not what we know. What is needed for us to know is unknown, and that we do not know this is also unknown—which is the biggest hurdle to escape the shadows of our action pathways. Escape is possible, but very unlikely, and here, changing (or not changing) geographies is the most promising of all escape strategies.

10 Fortuna

An imaginary lady, an all-controlling force from that there is no escaping. We like to refer to *Fortuna* in this way. But what is it really, what is behind such descriptions? Nietzsche (The tragi-comedy of Regensburg, 2014: 346–347) and others have placed great confidence in the importance of her. So, the question arises, are we being compelled to do the things we do, and in the ways we do them? What can, or possibly could, connect it all? Our decisions in the here and now, the decisions made in lost and forgotten empires long, long time ago.

The work of Fortuna appears as inter-complexities (as entanglements of entanglements), her ways of influencing/affecting everything. Everything is complex, everything is entangled, and yet everything complex is again connected in complex ways, i.e. everything entangled is once more entangled in more stealthily complex ways. This is her, Fortuna, the sum of great powers that arise out of overall (communicational) system complexity.

The flow of human communications, in the tube of history (our story), in the local sections in each geographical or communicational (internet, etc.) locality, brings about these great powers of Fortuna. They cannot be looked at and they cannot be touched with hands or tools (as in quantum physics). When we try to comprehend and reason them, we yet are running in these very wheels of Fortuna that we try to understand. That is, we run like mice in wheels that are connected with yet more, sheer endless wheels of human actions, human communications. So, it is not likely, but yet not impossible, just unlikely, that a mouse within these wheels of fortune, may grasp its running, the purpose of and the forces behind, plus the forces that have created and maintain the very same.

11 The Healing and Liberating Forces of Music

Our brain works in ways that are not easily understood. Basically, everything is connected to everything, and this is why certain things can and do exert greatest influence on everything. Music has healing powers a great many medicines cannot rival. It is stimulating, spreading joy, it is repeating, giving confidence. It is to be expected to bring more joy as the music goes on. Expectations, memories are all in the play. Large regions of the brain are activated and catered to, their joy accumulated and strengthening one another.

12 Deserts of Joy

When we see the flowers bloom in the desert, we are aware of the wonders in the world. In a figurative sense, these wonders may represent imaginary states of worlds that are there (that we generate) to make us believe things in front of us, in the here and now, or later on in our lives, are the things we want. Quite often this is not the case, as it turns out later. And we became victims of our own misguided believes, habits and rationale. Phantoms of misguided happiness are like drugs, they pull out happiness from within us while faking new happiness. These wrong kinds of happiness drain out true happiness from within our bodies, minds and souls. Money and wealth; career, power and fame; independence and security, they are all representatives of the *fata morganas* that we chase.

13 The Silos of Fortune

Fortune has more than one meaning, buying a flat in one of the sheer millions of high-rise silos of modern cities and especially in their suburbs carries with its very own mix of different kinds of fortunes, good ones and bad ones. These include far more than the eye can catch, or the financial status quo at any given time can tell.

These are not just silos of living rooms, kitchens and bed rooms, for the next generation to grow up, and the older generation to re*tire* into (after their *tire*some lives). These are also, and this is especially interesting here, silos of thoughts, silos of ideas and outlooks onto the world, and with it, all in all, silos of more or less common life-trajectories. The physical world has a most powerful grip on the psychological world, and the social and financial world. It is nowhere faster to be seen anywhere in the world (maybe as clear, but not as fast) than in China, where skyscrapers grow like mushrooms and millions of new fortune-seekers come to fill them as if this is a piece of cake, a super easy thing to do.

They think it was their decision, they think all will be good, and many times it is, were there not cultural changes, societal changes, economics changes, ecological changes which mix the cards of fortune again and again anew, and the hand one gets played changes constantly, sometimes too slowly that one does not notice it, sometimes with a big bang, and everything changes.

14 Time Is but Memory

The veins of our hands tell us time. The fossils in the stones tell us about their time. What will we tell, what will we be leaving behind? In the universe's time, we will have lived. Will we have lived with passion? Compassion amongst us, compassion with our planet, the earth? Perhaps the goal is to achieve a *perfect* earth, one that can go into the memory of the universe, by just having been there. Non-terrestrials, eons away, may find in time ways to look at us, as we live now, like we are acting in real-time, but long, long gone.

15 Empty Love

Not all things called love are love. There are plenty of shallow appearances of love, shallow practices of love. For many, love is a duty, a habit, a feeling of familiarity and, here and there, comfort.

16 Scary Love

Love is scary when it causes anxiety. It should not be, it should be a source of calm and strength, and so many other things. When the people you think that you love are anxious around you, afraid of you or outright scared of you, then you are not loving them and they are not being loved by you. Love of the past may have already for a longer while been transformed into domination, subordination and victimization of others.

17 Rationalizing and Killing of Love

Love can only be when there is care and respect, full care and respect, full dignity of others, and these are to be judged by the standards of those being (supposedly, or not) loved. Rationalization to one's own advantages is the enemy of love and relationships. Punishment of others with cold and angry looks, harsh and constantly repeated criticism, slamming doors and throwing things is the antipode of love. Love is usually diminished and then killed by the lack of positive non-verbal and verbal communications and presence of negative non-verbal and verbal communications. The build up to this, and all the way to the culmination of the killing of love, and long thereafter, is marked and pathed by myriads of rationalizations, in one-sided favor of one(self) and against the other, often reversing actual facts entirely, and twisting truths, said words and logic.

18 Container Competitions

People are so agitated, sometimes so sad, sometimes so enraged, by what happens in sport competitions. The root for the very most part for their national team, or their hometown team. When a small country, or town, loses against a big country or city, then the losing team's supporters are also, in the very most cases, beaten down, sad and angry. Euphoria, pride and narcissism is characterizing the emotional outcomes of supporters of the large country and city, while the reason for winning, logically, for the very most part, comes from the size of the country, the size of their hometown, or the money the football club owners provide for buying the most expensive players.

By the way, nationalism is a disease, it is a human and civilizational disease, not an achievement. When one looks at the countless wars and fights, the sheer endless anger and hate created just within e.g. football games alone, one may see that this is a mistake of epic proportions to think and act in categories, when e.g. looking at Ulrich Beck's *container theory* (Beck, 2006, 2007, 2009b, 2016; Beck & Sznaider, 2006). The greatest mental containers of all, nation states, are historically-determined on a random basis (of past battles and wars, and who had more soldiers, etc.).

From a universal point of view, I see no sporting spirit in a Dragon Boat competition between e.g. Mainland China, Hong Kong, Macao, and Cambodia. Too big is the one country, and too small the populations of the other regions and countries, no matter which countries, regions or cities we are talking about. The same is true in many Olympic disciplines. If the larger countries were to come with regional teams, e.g. on a provincial basis, or state-level or regional basis, then this would perhaps level the playing field. And in addition, and this is the key, we would hopefully finally come to a point where—*in general*—historical (ourstorical) containers, national flags and hymns are not needed, a video-introduction of their home province instead and their favorite song (of the players) instead would do.

19 Holding the Rail

Life needs guidance all the way, in all and/or many ways. Sometimes, this is meant literally. Growing a habit of holding the rail when walking the staircase, especially downwards, is better than a life-insurance policy. One can grow this habit, for example, thinking of one's favorite singer, or movie star, etc., who had to leave this world due to a tragic staircase accident, way too early needless to say. Every time when walking down the staircase, one can gently remind oneself of that favorite person, remember e.g. the cool concert or the good times one associates with that person, and every time one does so, one creates a stronger support for a positive habit, for now and especially for later in life, when it is needed most. It is training one's mindfulness, by being mindful.

20 Pear Candy

I only can guess how a pear candy causes excitement to the child that is sucking the candy. No thought about health or one's figure, that is for sure. Children have it easier to enjoy stuff. We adults have learned too much, to be possibly enjoying the combined flavors of pears and sugar.

21 Holidays

It is widely known and more widely still, experienced, all over the world. Holidays particularly are stressful, they bring out anger and hate, for all to see, and felt by all of those who are closed to the person in question, that is, especially partners and children, older parents (close family members). People let themselves go, people let themselves be harmful to others. They become emotional predators, as they do not fear repercussions or any other kind of liability for their actions.

They are assured of no consequences, as marriage and divorce laws and practices have '*locked in*' everyone: kids are 'locked in' and bound to certain parents, as well as all parents, they are also 'locked in' and bound to their children. The rules of the legislatures and the practices of the courts reflect societal myths and governing rationales, but they also are causing multiple damages on multiple levels in families, in society.

Threatened by powerful consequences for parental rights, caring needs and lifetime finances, the victims give in. The laws and practices of those in power add to the suffering, as they themselves are the source of fear and the frozen landscape of abuse and neglect, especially against the elder generation of older parents, but also against partners and children.

22 Towards a Theory of Human Entanglements: A Theory of Asymmetrical Human Inter-complexities

For many keywords, the normal meaning of the word is not the meaning of it when used in science. Complexity, in Niklas Luhmann's (1984, 2012a, b) theory world, means order, systems, institutions, policies, methods and strategies applied, particular solutions for particular problems and dangers. Of course, these are also all numerous and hence complex in the normal people's usage. When we, in social and human sciences, talk about complexity, we hence talk about the particular inner workings and inner systems that are operating through particular programs and plans, and particular modes of operations.

For Luhmann it was important to note that complexities of the inside world had to match complexities of the outside world, for the inside world to have a chance to survive in the long run. Hence, everybody and everything needs to adapt to his/her/their/its environment; every company needs to adapt to its competitors, changed customer preferences, new technologies, new regulations, and so forth.

Now taking this further, we need to investigate these systems and complexities made out of communications—of now all sorts of communications, in a post-Luhmannian (psychology- and biology-inclusive) tradition—and to explain them further. This is the task of social sciences and life sciences.

Michel Foucault (1975, 1976), building on and applying the essence of Nietzsche's key works (1878, 1882/1887, 1887), theorized and explained perfectly the workings of *power relations*. A large followership of modern-day social scientists has come to appreciate the usefulness of Foucault's *power relations*, that is power manifested in the human minds, which is power of some people over other people, certain groups of people over the rest of the population. Foucault (1975) left room for protest and revolution, but basically his system of 'civil war' is a system of 'elite war' fought and carried on by the elites against the rest of the society, from which the elites gain their privileges and one-sided advantages.

Power relations are manifested through language, education, culture, governmental institutions and such. They are forces and determining people's action, largely and mostly without them knowing of their existence, of their powerful effects and working mechanisms (through language, culture, education, media, etc.). People take local realities, their lifeworlds, for granted, the way things work, who has the say over what, and who has to obey, follow and swallow the rules, regulations, institutional disadvantages, punishments, stigma, and so forth.

Combining post-Luhmannian theory with Foucault's power relations theory (Foucault, 1975, 1976, 2009, 2010, 2014; Lukes, 2005, Rabinow, 2010), we arrive at a *theory of super entanglement*, a theory of inter-complexities, one that *explains the complexities* of one or other kind of personal and/or social thought-and-language systems, *and* more important, *their inter-complexities*, which are the result of Foucauldian power relations and the reasons why they exist in the first place.

Language created rules, stereotypes, meanings of good and bad, and *complex language constellations and long-term traditions of certain ways of using and directing the very same* have created power relations or, in other words, entanglements, which are in fact asymmetrical inter-complexities.

Entanglements are *inter-complexities that make things happen or not happen.* In the theory of human entanglements, or asymmetric human inter-complexities, the super-rich and the powerful, the economic, intellectual and political elites, are in position of creating, reinforcing and maintaining asymmetrical power relations to their very own advantage, and for the disadvantage of the rest of the population and to even let the rest of the population *pay* for maintaining and erecting these power relations (see Randall G. Holcombe's marvelous description of political capitalism, as he coined it, Holcombe, 2018, 2021, 2022; as well as my recent book on super inequality, Aspalter, 2023b).

Chapter 6
Bouncing to the Empirical, then Back to Quantum Theory Building

1 Neuro-linguistic Programming

Many have heard it, the abbreviation, NLP. It is simple, the gist of it (cf. e.g. Grinder & Pucelik, 2013). A myriad of ways to improve one's language developed over time, and have become part of the now very broad and inclusive concept of neuro-linguistic programming (cf. e.g. Tosey & Mathison, 2010; Boroditsky, 2009; Mathison, 2004). For me, over the years, I have picked out and grown some key findings.

The very most of words carry a positive or a negative meaning, or a tendency towards either. What we learn in school, is not how language is processed and understood by our human brain. Grammar does not penetrate the deepest depths of our brain. Our brain works with pictures, ideas and memories, not with grammar, it does not go with the official, textbook meaning of words either. Rather the brain works in simple ways putting words together or cutting them into its main constituent pieces. When the brain hears the word 'inexpensive' the first and lasting impression to come to mind, for the mind, is 'expensive'. Generally, negations are not heard by and focused on by the inner realms of the brain, the majority of the brain. If you want to say inexpensive, just say cheap, then the brain has got the message.

Words themselves, too, are received as pictures, much like the Chinese characters. The word for woman/female in Chinese is a picture of women kneeing down, the word for man/male in Chinese is a picture composed of the words for field and strength. History (our story) and traditional thinking are still very much alive in all of Chinese writing, as it is with all languages in general.

In any language, words are decomposed into their parts by the brain. Also, the brain calls up similar words and meanings, so one word seldom comes alone. The past and recent usages of words, in one's lifeworld and in one's particular language culture, also is of greatest importance. The word gay today has a very different meaning than what is used to be. If one would say the weather today is gay, most young people, and most people really, would not think of cheerful weather at all. The meaning of words can potentially change rather fast, but it can also reach far back

C. Aspalter, *Human Entanglement Theory*,
https://doi.org/10.1007/978-981-97-5161-7_6

into history/ourstory of the usage of this or that particular word. There is a myriad of examples for this.

Most people still think in their native dialect or native language, not in formal school language. In the local language or local dialect, things have more emotional meaning, a more close and familiar meaning and words stick to the mind much closer, much deeper and much longer, as formal language or school language would do (cf. the study of Jablonka et al., 2012).

So, one should use positive words, that contain positive, positive local, and appropriate contemporary meanings to trigger a positive reaction of the other side of the communication. This can be another person, a broader audience, or simply just oneself, when one is thinking, when one is remembering, when one is talking to oneself.

Language used and feelings triggered/recalled always are *programming* future human action and future human communications.

2 The Knee

Many people have them, a knee that is telling them things, different things. Some have a hip or a shoulder that is telling about a change in weather. Some have a knee or a foot that is complaining about stairs climbed. What they are really saying is take care of me, please take care. Pain is a reminder of doing things differently, of preventing current pain, as much as of warning about future changes.

Most people will put on some medicine, as they were told to do so, or as they remember TV commercials for doing the very same. Few will actually change their diet, and actually start to exercise, in whatever appropriate manner and ways possible. Symptoms do not need remedy, but the causes do. Hence, a changed diet and exercise go a long way in reducing, undoing and preventing pain.

Unfortunately, this message is not coming across, as it does not make billions of profits for multinational corporations, and for the health care system, that is either charging the patients directly or indirectly (including all other citizens) by way of social insurance reimbursements that come from mandatory social insurance contributions, plus other taxes.

Government officers, politicians do only care about their salaries and post-political-office careers (cf. esp. e.g. Niskanen, 1975). Many of them are also not informed, about chronic inflammation and what it does to the human body, from arthritis, all stages of diabetes, cardiovascular diseases and, of course, all the way to cancer. The universities long for research money, for the purpose of getting research money, not for the purpose of rescuing society from knee pain, arthritis and disability; or diabetes, heart attack and cancer. Hence, science, too, is compromised.

And the poor knee keeps on hurting and hurting, and money keeps on being spent on the wrong things, medicines instead of apples, bananas, oranges and jogging shoes.

3 Only in One's Thoughts One Is Free, or not?

If the human mind is a place where thinking can lead to enlightenment, Buddha (Siddhartha Gautama) would have gone into a library to get enlightened. The fact that he went into the woods, to be sitting under a tree, tells a tale. Thoughts are the enemy of enlightenment; rational thinking, too. Thoughts are in fact the prisons we live all our lives in, with barely anyone making a prison escape. This is where Siddhartha (Hesse, 1922) is coming in, he broke out. By not exiting the prison, but by expelling the prison from the brain, by expelling thoughts, notions and mental formations (ideas, hate, etc.).

Gustave Le Bon (1881, 1894, 1895, 1918) and Sigmund Freud (1921) were among those that were very clear about the impact of society on the inner workings of the brain, where all of society's rules, dos and don'ts, all of its evaluations, all of what is good and bad, and why, are embodied in the super ego. Lev Vygotsky (1978), too, perfectly understood and perfectly explained the causal relationship between a person's social environment, and not just parents and family, and the development of the human brain, the human character and lifetime behavior.

4 Why Did They Burn Down the Town Every 40 Years?

Today, science does not have all the answers. The first high civilization in Europe, the first one on earth that used writing symbols (that we know of today), the Vinca are giving one particularly interesting puzzle, or case study, for us to ponder about (cf. Gimbutas, 1973, 2001; Chapman, 1981). They were a metal-working civilization, with high specialization. They lived on the river Danube, in and around modern-day Serbia. They were mostly a female-dominated society, as most early societies were, from Europe to the Indus Valley, due to the nature of hunting versus goat and sheep farming at that time. The Vinca were peaceful, with no war faring, no emphasis on weapon manufacturing. The Vinca shared all their food, and stored them in central storages in town. But one thing is peculiar, they burned the town regularly to the ground, every 40 years. Was it religion? The need to prevent diseases? Probably the former, as in the latter case the exact number of years, 40, does not make sense. It was not Keynesian labor market policy (cf. Keynes, 1936), again why exactly 40 years, again and again. Wooden houses easily can last for 200 years and much longer. So why? Ideas must have been in play. Ideas, fears or elite/political calculation.

5 Plastic Straws and Plastic Gloves

The world would be so much better a place, if politicians would forbid, as simply as that, all plastic straws and plastic gloves, all one-time-use plastic utensils. And they for the very most part don't, they have waited for many decades. Inertia and procrastination, neglect and selfishness, all are to blame. Politics does not work, as markets do not work, in many cases—in many most important cases this is the case. The reason is to be found when following the money.

As long as billions of dollars or Euros are made, politicians side with the manufacturers and fast-food empires, et cetera, that are using these horrible plastic items that pollute all of our earth, all oceans and all glaciers (those that are still left, for now), politicians will side with them, by not doing anything that really would stop the madness. Empty talks and treaties, all but for political show and personal self-esteem. Nothing is as easy as forbidding production, storage and usage of these plastic items.

Consumers are also to blame, as they in some parts of the world like to be 'comfortable', 'clean', 'high-class', 'developed' and such, which is all rationalizing non-sense of course. But they build these rationales and defend their usage of such items with egoistic non-sense, shutting out the essential larger picture. Many of them actually never thought about it, as they just use what is put in front of them.

It is a culture of non-culture, a culture of rationalized recklessness and rationalized cold-bloodedness.

6 The Power of Repetition

Human programming needs repetition, it is the best way to softly program ourselves, or others. Also, multiple angles of one and the very same idea or message are deepening and strengthening this or that idea, reasoning and, then, behavior or non-behavior. Mountains can be moved with repeating words, with actual or seemingly/pretentiously logical arguments and personal/social/political goals. But with them, also cities and entire countries can be bombed or burned to the ground!

7 Crazy Ideas, Ingenious Ideas

Are humans capable of quantum computing in their brains? That is, on a *quantum physics* level. Some scientists strongly believe so (not me), and they believe they have worked out first theoretical ways to test their theory. This is (generally) how humanity, civilization and society, with all of its knowledge, science, is progressing. Crazy ideas, some succeed and change the world, and some not. Creativity is vital for deep and far ventures into the universe of the unknown.

8 With Time

In the evolution of all-in human communication, thoughts and feelings included, and hence the evolution of human action, everything is dependent on the past. Everything forward from now is depending on its past. As our local (earth-bound) time moves on, the entire tube of human (all-encompassing) communications is progressing. And as it does, it also enhances its complexity, as the size of communicators is growing, as their frequency of interactive (and hence non-interactive) communications is heightening, as new ways of and new technologies used for communications are invented and developed.

Human inter-complexities, that is, complexities of complexities, connections in between complexities are also increasing with time passing on, from the pre-human, early human times into the future. The number, mechanisms and effects of human inter-complexities are increasing. This does not say however anything about the nature of these inter-complexities.

Human entanglements are particular human inter-complexities, they exist where one system of communications is impeding and compelling communications and actions of certain communicators, groups and strata of communicators that are outside the controlling/impeding system, and far away from the levers of power of this system of communications.

Human entanglements are created by establishment, maintenance and promotion of (often dichotomic) evaluations, legitimizing arguments and rationales, and of course attempts to convince (or hijack) others to support rules, rulings, laws, policies and practices (cf. e.g. Gramsci, 2011a–c; Habermas, 1981a: 39).

These are pre-authorized—non-self-authorized, non-questioned—common sense arguments, applied/manufactured logics, rules and evaluations. These entangling systems hide in secrecy and complexity. They cover up with denial, confusion, lies and confrontation (cf. also Plato, 2011: 514a–521d).

With time, the ferociousness of human entanglements increases, also their stealth. With every countermeasure, new measures of undoing, going around, off-setting the very same are established—in many places, from different actors, and in silent coordination, as the elites are *all in*, when it comes to profit from exploitation of the masses, from the disadvantaging of the masses, from the self-advantaging of the elites (monopolies, exclusive access to finance and power, and so forth) (cf. ibid., as well as Holcombe, 2018, 2021, 2022).

Chapter 7
Down to Earth Again—Staying in Touch with What Matters: The Empirical World

1 Urban Disneyland

It matters so much where one is, in a village, or in the city. We are used to it, the craze of noises, the maze of building, speeding cars, redlights turning streets into organized conveyor belts. Trees and parks are decorations for the roads. Flowers are planted at the crossroads to maximize their consumption, looks received by the bypassing road users (mostly or only). Monotone skyscrapers disguised as towers of glory. Hospitals and mid- and higher-level schools to be proud of. Factories are the running engines of it all. Stranger things there are, yet it is pretty fantastic: the admiration and the praise and gaze of the ordinary town and city folks, for their engineered, commercialized and *highly controlled* wonderland.

2 Our Picture of the Neanderthals

We are so full of ourselves that we had to, and still have to, see Neanderthals as wild beasts, as opposites to the perceived ourself. But, actually, it was them, the Neanderthals who also invented/reinvented art/music (flute, etc.), religion (shamanism, its predecessors of course, comes to mind) and with that funerals, one of the first welfare institutions of all. The Neanderthals are painted as losers, as vanished, and yet most Europeans, Middle Easterners, Asians and most inhabitants of the modern-day Americas and Down Under carry their DNA. Asians also carry Denisovan DNA, a sister group of the Neanderthals (cf. e.g. Pääbo, 2015). Thus, one cannot say Neanderthals, and Denisovans, are not successful; well, they were not completely successful either. Are we, will we be?

C. Aspalter, *Human Entanglement Theory*,
https://doi.org/10.1007/978-981-97-5161-7_7

3 The Broken Narrow Streets of Daibo

Cruising slowly through the narrow streets of Daibo, I am not upset about the bumpy underground, broken concrete, holes in the road, hastily and most cheaply repaired roads. They are a safety feature; they are keeping the town and its controlling enforcers out. These hence are concentrating their efforts of controlling and punishing/fining people (usually due to the usage of non-authorized/illegal motorcycles, not wearing helmets, or selling vegetables on the side of the road, etc.) to the larger entrances, the main gates to the village, which is a village inside the town, as there are everywhere all across this country, also in the biggest cities.

4 The Defiant Ones

It does amaze me quite so often, how often people are risking their lives for two seconds or a half-a-minute 'gain' on the road when they run a red light. This is very common here. Maybe they have important things to do, they for sure are hard working all the time.

People like these in general can be thought of as non-100%-'civilized' (as people in general understand it) or better '*not 100% molded*' (as I understand it)—which at a closer look I find intriguing and encouraging indeed. Sometimes, so it happens, they are not controlled, despite all the controls in place. They are defiant as can be, when looking at them from a docile Alpine/Middle European perspective (cf. e.g. Ringel, 2005).

They may, for some, serve as role models for anyone in the world who wants (or would want) to learn to be, and how to be, defiant. So, when one—as a Middle European—is looking at past pictures of a person standing in front of a military tank, with bags full of vegetables from the market, this does not (and would not) surprise me. That is, or would be, just another ordinary person. As, here in this part of the world (all of Asia), defiance is (still!) an art, a culture, which truly lets me be positive about avoiding too much of any kind of Orwellian future (Orwell, 1949), for *any* society on earth, for as long as there are defiant people like that. So the world is better off with all its diversity in play.

On can say, this is against these or those principles, we need to be more orderly. But, then, what is the nature of humanity? The nature of freedom? *To everyone as they wish.* Romans do it Roman style. Indians do it Indian style. It is the same all over the world, people have their own recipes, their own approach to do things. Not anyone needs to copy anyone. All as they wish.

Diversity is key, diversity is the secret to development, to a better future world, no matter how special (weird, or 'perverse'—i.e. *against the stream*) it may seem at first, any kind of diversity is beneficial for the advancement and betterment of humanity in the long run.

5 The Anthropocene

It is not entirely clear when it started, but it is clear that it started, and what is happening to the earth. A new epoch, that of the Anthropocene, has started. Be it with deforestation during the Roman times and the Middle Ages, the start of the Industrial Revolution with fossil burning on massive scale, human-made atomic radiation all over the world, or global micro-plastic pollution since the 1940s, our world has changed and is changing (cf. e.g. NG, 2023; Steffen, 2021). A new mass extinction of species, a new era of global warming, that is our record, it is a record of shame.

6 Daily Routines

It is rather well known that daily routines give comfort and a sense of security, they also give joy and happiness. As they are repeating, once one enjoys the fact that they can be enjoyed again, many times just knowing what is coming is already making one happy. It does not matter if these are work routines, health or happiness routines, they all support a regiment of security and help shaping and guiding one's overall and long-term trajectories. Mental health outcomes also largely depend on such daily routines, they are important for us, more than many realize or care to admit.

7 Bambini

We have managed to kill the feeling of children being a blessing in the Western world. There are exceptions, Italy is still a country where people are so much in love with kids, especially smaller and small kids, and babies of course. But this does not mean that in Italy many people have babies, or two babies at least (one for each parent). Responsibilities are out, lifelong partnerships (marriages, etc.) are out. So, children, bambini, are quickly becoming a thing of the past, not just one's own past, but the past of many societies.

There will be children, but way too less. Too less to carry on traditions, and societal peculiarities. Many cultures will blend and/or perish. Societies will blend and/or perish.

It is not hip to speak out the truth about the relative non-continuation of Western societies and cultures. But yet, here we are in a situation where having kids is not cool anymore; mobbing and bullying in form of medial assaults of all kinds (e.g. assaults in movies, in magazines, and especially assaults in all forms of social media) have risen, and the passion and rational support, as well as practical support—as social security is provided by institutions and not children anymore—have long vanished or become out of fashion, a thing not to do, not to desire.

The modern welfare state and modern cultures of this and that have killed societal fertility and with it societal happiness, all in one.

8 The Story About Adam and Eve: In Essence a True Story

Modern and ancient civilizations are and were marked by it, by long-gone and still long-felt events that are vital for the understanding of history (our story) of humankind, as well as this or that religious fairytale. One such fairytale is based on true events and actually altered human history, our story, a great deal. This is the story of Adam and Eve.

It is a close-to-reality story, not names of these persons, not these persons are important, but what they, people like them, did (not literally, but still), what happened, what had changed their world at that time. We have today a pretty clear idea when and where the events that have built this story happened. This happened about a bit more than 11,000 years ago, the place is today called *Göpekli Tepe* and lays in modern-day southern Anatolia, in Türkiye, close to the border of Iraq (Collins, 2014; Seewald, 2020).

The going and most explanatory theory today is that this site is a giant BBQ and festival site, much like modern-day Oktoberfest *Wiesn* in *Minga* (a.k.a. Munich), again not literally, but much like it. Bones of animals were scattered all over the place, rather much of the meat was still on them. So, there was plenty of BBQ meat to go by. But, even more important are the big pots with traces of beer that have been found at that site. And, and this is the key, in order to make beer one needed crops, in order to have crops, one needed to plough fields, and tend the fields, which is exactly the punishment for Adam and Eve (i.e. people in general, not two specific people), after they (not ate an apple, but) started to party.

BBQ and beer that is a party, pretty much just like the Oktoberfest. Each table had columns with different logos or signs, much like theme tables, that belonged to different groups or families (much like exclusive *Stammtische*, or regulars' tables, one can assume). Once you make a little bit of beer, and get used to party, Octoberfest style, then you need to make more beer. Beer was difficult to make, meat hunting was relatively easy and quick.

From about 10,500 years ago, agriculture was developed in *Göpekli Tepe* (cf. ibid.), this marked the start of the human agricultural revolution, and the beginning of the second great migration from modern-day southern Anatolia and northwestern Iran, including the Northern Levante area, to Europe, all the way to Norway, Scotland and Portugal.

So, for sure, this was a long-term process and development, but the gist of the story is valid. Life in agriculture was marked by strenuous labor, and high risks of failed harvests. This is the story of longer institutionalized BBQ/hunting seasons and beer 'festivals', and *not* of an apple tree and a snake.

Religion had to edit the story, to make it more tellable; religion had (could not refrain/resist) to add a greatly exaggerated moralizing content, with a much scarier content (especially for kids), with the intent of influencing and controlling people.

All in all, we can say archeology, and science in general, are great. This is a good story, a good story to tell, and a good story to place in context our modern-day societies and modern-day development mania. We are not so much different from the people 11,000 years ago, actually pretty much the same, not much has changed, apart from technology of course.

Chapter 8
The Quantum Methodology Matures

1 From Archaeology and Genealogy to Quantumology

Foucault (1954, 1962, 1963, 1969, 1975) is most-known for his methodology to use the study of history, of the history of thought systems to be precise, in order to analyze modern-day societies. He delivered a most descriptive and most explanatory account, of how people are like puppets of their own, as the strings that hold up and push the puppet are going back into the history (our story) of one's personal, one's family, one's culture and one's language tradition. Nietzsche was the grandmaster who taught Foucault the basic outline and the methodology. Foucault (1963, 1975) applied it mostly to the late medieval times, but also to the classical ancient and early modern times. Nietzsche (1887) even went back thousands of years, to analyze the now and here.

At first, Foucault called his approach the archeology method, digging out meaning and reasons for existence and change, reasons for behavior and developments, from ancient texts and stories, from the study of language and from history (ourstory) itself. Then, Foucault developed his method further and called it also genealogy. Both ideas for method development are directly derived from Nietzsche's *On the Genealogy of Morals* book (1887).

In our book here, we have developed a different approach to analyzing any historical, social, psychological, or linguistic research object. Due to the immense complexity of the research object under scrutiny, the entirety of human communications, including intra- and interpersonal communications, from the beginnings of early humans up to this very moment in time, we apply a different method than archeology or genealogy. It is a spin off from the two, for a different purpose, and it is different in its setup and execution. This method that has its very own characteristics can be called *quantumology*, as it applies a quantum-theory-inspired approach, which can be employed in all of humanities, social sciences and beyond.

The key essence and differences to other methods are the emphasis on (1) randomness (as much as possible), plus (2) the switching in between levels: (i) theory and

C. Aspalter, *Human Entanglement Theory*,
https://doi.org/10.1007/978-981-97-5161-7_8

empirical levels, (ii) meta, macro, meso, micro and micro-micro theory and observation levels, (iii) from the inductive to the deductive in an alternating manner (applying an intrinsically abductive method), (iv) all the way back and forth in history (our story), (v) in and out of different scientific disciplines (i.e. it is truly transdisciplinary). A key feature of quantum computing (3) is working in parallel, i.e. analyzing, in this case (here), different perspectives (dimensions) of human entanglements. In addition, (4) it works with both logic and pondering, to create and to innovate at the same time (in the same research study, or research program). Also, (5) it connects all—useful/powerful—previous theories and (6) tries to connect all kinds of empirical evidence, such as observations, and the current state of the art of science.

This method on purpose is emphasizing randomness, plus constantly changing perspectives/angles of observation, in order to cover (as much as possible) everything, to keep an open eye for everything, to connect everything, and to double-check everything with everything.

For randomness we, here, are making sure that we are catching the average as well as the marginal phenomena, and thus are getting all kinds of data/input. In addition, we are constantly changing perspectives/angles of observation, in order to achieve higher levels of triangulation, due to the special nature of universal, even high-dimensional, communications:

> e.g. suppression of a feeling or a revolution on the street, is always suppression, a inhibiting force; poverty of emotions and thoughts, is lack of resources, just as financial and cultural poverty.

The method applied in this book copies the idea of using *multi-dimensional qudits in quantum computing*. It is necessary to understand that *qudits* feature multiple layers or dimensions of expressions, with data being stored and read differently, to increase complexity and hence computing power. In contrast, *qubits* measure only two different states/outcome, after the state of superposition.

Here, this is achieved by switching from induction—i.e. inference from observations and other empirical evidence—to deduction—i.e. inference from theory and logic; from prehistory (pre-our-story) to current times; from the macro levels of governance and culture to the micro levels of human actions, human communications, human thinking and feelings; from developing theory to developing methodology; and vice versa.

As with quantum computing, it is important to preserve *universality* of gates (i.e., quantum gates), or transformations points (cf e.g. Fouché et al., 2008). That is to say, we need the same base units, we need to be able to match everything analytically, and this we can, by using *communication* as our means of interest, our all-connecting thread. Language (feelings are and can also be expressed in words/language) is our uniting dimension that unites all research observations, all random phenomena that have been *in bulk* put under the microscope. In other words, we are operating solely in the communication universe and in all its corners and aspects, at all its levels.

As with quantum computing (Hou, 2023), we here are investigating (operating, computing) *in bulk, in parallel*. The importance of which cannot be overstressed. Operating in parallel is a core qualifying characteristic of quantum computing.

In this way, we can harvest and compute (work with) information (knowledge) that otherwise would not be possible, i.e. not achievable—not in the here-applied form, with its special characteristics, analyzing and explaining a myriad of things (entanglements, interactions) at the same time.

The purpose is to create knowledge about forces that are otherwise not directly felt and not to be seen, they cannot be touched and *easily* imagined, even though they operate in plain sight, always have and always will.

These are *the forces of human entanglement* that operate through all forms of human communications. These are *asymmetrical inter-complexities*, in the Luhmannian and Foucauldian traditions of the meaning of complexities and asymmetrical power relations respectively.

2 Diggers

Many kids love them. As it turns out they are everywhere, in this coastal part of this industrial giant, where everything gets built, everywhere. When I think, I dig. When I teach, speak or write, I dig. Into what? The only answer to that, that can be, is: everything.

It is like the internet, working on those transcontinental cables that facilitate worldwide communications. You dig in there and if you are lucky, or try to look hard and long (and skillfully, methodologically skillfully) you may get a great harvest, at least over time. And time it took to dig out and assemble theories, like a (in the double sense of the word) *great* puzzle. It is clear now, clear enough, for me.

The *Theory of Human Entanglement* has emerged and stands on solid ground and it can fulfill its function to serve knowledge creation (science). Let's see later how much more there is yet to be discovered and brought to light, once the silent veil of publication has been lifted, and the responses are coming in.

3 Nudging

Some love it, this new buzzword, and others have found it to be quite (or should I say rather a lot or completely) unethical, that is, to push others to do what one wants them to do is robbing them of their choices, of their freedom of self-determination. It is on top bothering them, it is bullying them. So, how can that *not* be unethical, nudging people?

All are nudging now, pushing others to click on things, agree on things, they do not want to click and agree on. Governments and technology should stay in their lane and help people, and not push and bully people, to turn them into their marionettes of all sorts.

People need to be free, humanity needs to be free, free from unfreedom, free from pressure, free from predetermination and free from determination by others. This is a fundamental nature, aspect and need of humanity (cf Fromm, 1941; Aspalter, 2006). How can science and philosophy help? Unveil the monster!

And this particularly applies to countries that think of them to be rather different from what their common lower-class members actually experience in their daily lives, and all their lives long (the countries of the West). As Johann Wolfgang von Goethe noted:

> "[n]o one is more hopelessly enslaved than those who falsely believe to be free" (a *popular* translation from J.W. von Goethe, 1809: Chapter 5, Aus Ottiliens Tagebuche).

4 Birthdays

The older we get the more people think of us as a burden. But this *does not make sense.* As we celebrate our each and every birthday, we are celebrating the positive and wonderful event of having lived another whole year.

By the way, it is commonly understood that the young generation now, needs the support of the older generations (at least two of them). Yet, the young are not the carers, the carriers of the social security system, it is still the older mid-age generation. And all will depend on them. The mobile phone-addicted generations (more than one already) are the ones that need care, financially, emotionally, and yes medically. It seems more young people need the services of mental health hospitals than any older generation.

Getting older is a great achievement, for one personally and for society as a whole. For those who do not know it yet, catch up, with logic, and general statistics as well. Still, the younger generations wait and ask how much longer they have to wait to get their inheritance (*literally* there are cases like that, even with their parents still in their forties).

The older one gets, in fact, the less burden one is on the system, on social security, the health care system (beyond 75 years of age hospitals do not do much anymore, no operations, much less medicines, too dangerous for them/the patients). Hence, the older one gets, the less burden one is on government finances, comparatively speaking, if one adds up and compare all expenditures (per person) for the very old and for those who are not yet.

The very old, 85/90 and plus, are the cheapest and best citizens, as they have, by the very fact that they lived to such a high age, proven to have lived healthily—and thus cheaply, with no burden whatsoever for anyone; in terms of health care and in terms of very-expensive long-term care (cf e.g. Li et al., 2023).

They are not a burden. But quite the opposite is the case, as it is them (the ones with the highest age, according to their birth certificate, not necessarily according to their actual biological fitness and robustness, which may point to a relatively younger or much younger age) who are and have been carrying the burden of others, the burden of the younger unhealthy ones, the unhealthy young ones that do not get to see their much older days. The older people of today, in most developed and developing countries give away lots of trillions of dollars (Euros, etc.) to the younger generation every year. Their pensions are functioning as child support systems, their adult children included.

Societies, the media, governments and intellectuals, they are all programmed to repeat after one another. They seldom fully, really deeply think, think it all. They are busy buzzing around their whole life, trying to get ahead of where they are, not even trying the see the whole picture, the evidence and the greater picture.

Old age is a blessing, it is a secret, a success story, for the people who make it to that high age and for society, which should learn from them and be really nice to them. That is not difficult, is it?

5 Horizons

As far as the eye can see, it is a feeling, a notion, a wonderful exciting and romantic concept. Adventures are awaiting on the other side of these horizons. The seafarers knew, they took immeasurable (for them, at that time) risks to embark on the most challenging journeys of all. Some of these journeys really paid off greatly. Some explorers changed history (our story), for the better, as much as for much worse (e.g. Cristoforo Colombo, a.k.a. Christopher Columbus). Many others did not live to tell the tale (e.g. Fernão de Magalhães, a.k.a. Ferdinand Magellan). It changed us, perhaps even just by looking at these horizons.

In the future, humankind will embark on great adventures outside of our earth, and even our solar system. Many will not make it, and a few will change our story then, forever.

6 Boredom

Some say there is nothing more dangerous than boredom. Many fatalities of young people support this. Some say people, those who are smart tend to get bored easily. Others say the opposite.

Take playing cards and playing mahjong, for example, it is all about killing time. It is about not knowing what else to do, to do with the time on hand. One could say these people are not bored because they play these games, or one could say they are that is why they play these games.

There are scores of people who are so busy working for their families, for their kids, that they have no time to even think about boredom. Different people are so different from one another, we all live in our own context, dealing with our own life conditions—in our own corners, in our limited micro universes of our own.

7 Human Greed and Folly

The Tulip Mania of 1636 and 1637 in Holland, or the Wall Street Crash of 1929, and many more of such events are all witness to what is driving our societies, day in and day out. The want for more, the blinding craze of wanting more is pushing us and pulling us like little leaves out there on the stormy ocean.

Not all people are on that train, many never had a chance to get on it, many never wanted to, and others luckily avoided getting on the very same.

Chapter 9
Analytical Reflections: Human Entanglement Theory Applies to All Aspects of Human Lives and Civilization

1 From Peruvian Roads to World Hunger

Every so often, governments and societies find themselves in a state of being twisted. They are working against the well-being of the society, the people. Procrastination and absurdity, carelessness and corruption, all have the same outcomes, things are not done, policies are not made, or things are done poorly, or very belated. The unbuilt roads of Peru (*Medinaceli Baldivieso, 2017*), or anywhere in the world, the unbuilt schools and hospitals anywhere in the world, as well as world hunger, and so many more of our shortcomings are rooted in the problem of politics (not just a political problem, politics is the problem), but they are also rooted in fundamental human characteristics.

Looking at developed Western countries, public choice theory in the past has named the problem, people in government, at the levers of power, are not interested in helping the country, the world's population, the planet. They just care for themselves (Niskanen, 1968, 1971, 1975, 1994; Buchanan et al., 1980; Buchanan, 2009). That explanation goes a long way (cf e.g. Ertürk & Whittle, 2015).

2 The Human Tragedy of Mass Incarcerations

The United States of America is one of the worst mass-incarcerating countries in the world, per population (cf e.g. Thompson, 2010; Raphael and Stoll, 2013; Stuart, 2018; Alexander, 2020; Longhi, 2023; Sawyer & Wagner, 2023). Why is a democracy fighting a war against its own people? Why aren't many people talking about it? The latter question seems to be even more intriguing. The first question's answer is easy, extreme capitalism is not the friend of the lower classes, nor that of the middle class either (at closer-up inspection). The latter question has to do with the upper middle class and the upper class controlling the media, academia and, of course,

C. Aspalter, *Human Entanglement Theory*,
https://doi.org/10.1007/978-981-97-5161-7_9

the government, the justice system, the legislature and all other levels and segments of government administrations. The system of mass incarceration works in favor of cutting job competition for the middle classes and lowering the position of the working classes, in favor of creating a cheap workforce for the capital-owning class, dividing the political electorate (making a united front against the rich, against the powerful, not possible or likely at all), shutting out large parts of the working classes from the electorate, and keeping the rest of society, the middle and working classes, in fear and full of anxiety.

Norway has another, a much more decent approach to offer (BPR, 2022). Why only a few, outside of Norway, and Scandinavia perhaps, are talking about Norway's system of incarceration that offers humanity, and works with humanity to make society a better place, right there with and in the justice system? Why the quiet?

The super-rich and the powerful, the intellectual and media elites have no interest in working for the favor of the poor, the working classes. The opposite is the case (cf e.g. Kreiß, 2023, 2019, 2015, 2014, 2013), they are quite content with how the system works in most countries around the world, as they are not doing anything against it, not even talking, and not even trying to replace it.

3 Real-Time Thoughts

Most thoughts never leave the realm of the person that created them, and so are nearly all feelings and memories. Private is what can be, for sure, definitely, locked in; locking out others who then cannot invade. Luhmann (1984, 2012a, 2012b) was clear about that. Social media, any play on the phone, is not private, as the cloud and anyone, any machine and government institution, etc., that wants to, can enter and join the room, read, watch/listen to, copy and store *everything* what is supposed to be, and mistaken to be, *private*, a.k.a personal.

Nietzsche (cf e.g. 2009b: 74–75), too, mentioned that very private thoughts, most inner thoughts, the many frequent, seemingly chaotic and mixed-up thoughts of any human are not to be shared with others, not involuntarily and not voluntarily either.

4 Who Remembers Yue Yue?

It was the year 2011, and all of China talked about it, and many in the world heard about it. A little girl was killed by being run over by two vans, by one and then by another one time after the other, and then being ignored by many people passing by. Time and again people passed by and no one helped, while she was still alive DM, 2011; as well as e.g. FlorCruz, 2011). Today's China is a different China, very different from the year 2011.

Back then, the incident hurt the entire nation. The government responded quietly by putting up more than half a billion traffic cameras within the following ten years.

All of China, as a consequence, changed. Influencing people's self-control was never sought, the authorities right away jumped to use technology to surveille traffic (and more), and to change behavior in this way.

The killing of Yue Yue, a two-year old girl hence has sped up the industrial surveillance of an entire country, and one fifth of the world's population. Tens of thousands of lives have been saved ever since, due to the changed mass behavior of traffic participants because of those many millions of cameras in place.

Now, I am not sure how many remember the little girl, her suffering, the heinous behavior and non-behavior of others, those who run her over, and driving off, knowing they run over a girl, like she was nothing, and those who walked by thought of her as if she were nothing. The psychology of a whole nation needed healing, but instead got a technocratically administered technical solution. More was needed then, more is needed now. Public psychology (societal psychology) is needed for many countries and regions, in normal dire times, in special dire times.

China indeed has advanced/progressed a great deal, and is now very much using mental health hospitals to help its younger generation, which was unthinkable back in 2011. China has changed a lot (it simply had to), with great achievements in a number of areas, not just in and with technology, especially in terms of economic development, but it also changed—and perhaps not noticed by the rest of the world—its attitudes towards mass-use of social work and psychiatry, remarkably and seemingly luckily so (because it had to).

5 Krishna to Arjuna

We need to stay in our lanes, work through live from and in our position, with the powers of what we got. Of course, we can get new powers, knowledge, or wisdom, technical skills, other skills. We can move to other continents, to a big city, change the most forming, form-giving of all things, our environment.

In the Bhagavad Gita, Krishna tells Arjuna, his friend and student, it is ok to do what one does, one needs to keep doing it. That is, not everyone can run away into the forests, the jungle and sit under a tree, for as long as it takes. Another key point from Krishna is all people can seek enlightenment, practice meditation and mindfulness. We need to be content, because with no contentment how can there be happiness and wisdom? But, this does not mean that everything should stay the same. The whole book, the Bhagavad Gita (Swami Prabhupada, 1968), is about change and transformation of the self, of the human mind, body and soul.

6 Nietzsche to Goethe

Goethe was held in highest esteem by Nietzsche, and many others for centuries. And both are now still. Nietzsche was sending Goethe a message back in time, in The Gay Science (2009a: 249).

Nietzsche hinted, in a small song/poem, that the all-ruling God has ceased to influence every aspect of all people's lives—and now someone else, the elites have replaced God. The elites control society now directly, not indirectly through God, whom they created, and through whom they used to control others, when and what they do, and why.

The smarter observants (the few lonely intellectuals, here and there) are calling this *fate*, which seems like history and time itself tell us what to do (where, how, and why), but in essence it is now the ruling class telling everyone else what to do, now in a direct manner, without the go-around of using God, or religion as an excuse, and disguise. The big masses are calling it *play*, and see this as part of playing *the game.* Now we could say following and chasing the American Dream, or some ideas and notions just like that. Both (the intellectuals, and the masses) are not seeing what is really going on, both are blinded.

This also very fittingly describes modern-day societies in large parts of the world, particularly the more developed parts that have lost faith, lost their old ways of thinking and rationalizing. But this is not the point. The key is people are the new Gods, the super-rich and the powerful are the now self-proclaimed Gods of modern times. It is not that they would say it, or emphasize it; they just act like it and hand out rules and tell everyone else what, when, where, how and why to do and not do all the things in life they do, want to do, or not.

The new Gods are stealthily being Gods, behind the scenes, behind the curtains, and starting centuries ago; there they have rigged the game, and ever more so are rigging the game, the conditions, the rules, the access to and exclusion from opportunities and fortunes. They control everything and let us pay for the system of controlling everything. Taxes, contributions, fees, cold tax progression, devaluation of salaries and pensions, punishments, court rulings, all of these are put in place, maintained and extended constantly, to pull down ordinary people, to make them weak, to give in, to give up, and say 'how high should I jump?' It truly is *the biggest scheme of all* (cf Aspalter, 2023b; Holcombe, 2018).

This is what one *can* read out of the backward-in-time communication between Nietzsche and Goethe (Nietzsche, 2009a: 249). Everyone can have his/her take on it. Communication is and becomes real, as long as it is understood this way. People long after their death can talk. Contexts can change, language, language usage, and their connotations are changing; but ideas can live on, and these ideas can create new ideas. There is an endless birth-to-birth cycle in the human noodle of communication—that started millions of years ago, and goes into the future millions of years more (we hope).

7 Gendering Other People's Gender: The Gender Entanglement

The world has changed, again, as it does so often in so many regards. Gender is not binary anymore; it is not to be perceived as binary anymore. It always was as it was, diverse, as diversity nature has provided for.

A couple of decades of trying to counterplay one gender against another was based fundamentally on the binary concept of gender. This decades-long fight was trying to achieve gender equality, by assuming the gender of others, and denying, at the same time, the gender of yet again others. Gender equality was the goal, and gender attack, gender molding, gender-forcing was big part of the outcome. But things have been changing (cf e.g. Bacchi & Eveline, 2010).

Now there has come a point, where even the greatest (fiercest) warriors of that form of 'the binary equality fight' have come to see and admit that the reality is like this: there are more than two genders, and one has not the right to assume the gender of the other.

That is exactly what happened in the last couple of decades, blaming others for being (assumingly) members of one particular gender group, which is a practice of genderism, and, yes, another form of sexism (if one looks the problem straight into the eye). There is, today, no more definition what a woman, or a man, is, who is this or that. People decide for themselves, what they want to be, who they want to be, and this is good so, and that was long, long overdue. It is no one's business to assume and blame others (directly or indirectly) for their accepted—or not—gender identity or sexuality.

8 The Might of the People

It is already a long time since the Berlin Wall was brought to fall. People are powerful if they want to be, if they need to be. All the powers of the earth are waiting to be released, when one part of the people is oppressing the majority of the population. It is wise to help the majority, to cater to the needs of the majority, and not to dictate, trying to mold and fold, and to suppress it with whatever means available and perceivable. The struggle for freedom is innate to the will and soul of the people (e.g. Flew, 1991; De Dijn, 2020; Davison, 2023).

The sad story all across the world that repeats itself so frequently, over shorter and much longer periods, is that former freedom fighters, former fighters for people's emancipation and liberation, become the new suppressors, dictators and tyrants of tomorrow. That is best described by Karl Polanyi's (1957) idea of a *double movement*, one or two steps up, then one or two steps down.

Christianity and socialism, and capitalism too, have failed to liberate people, they have all but stifled and smothered the ones they have, or supposedly have, liberated. They all tell people what to do, in the ever most comprehensive ways, they do not think people are free and mature individuals that can determine their own lives and fates in life. They all treat their citizens like unknowing and irresponsible kind of children.

A new look at how things are done and are not done is needed. A new outlook and a new paradigm of governance is in demand, who wants to deliver, who wants to help?

9 The Wonderful Nature of Literature

It is great, literature has the power to take us out on a ride, to take us away from sorrow and a monotone life. The best thinkers often are literary minds.

So many facets and levels of the inter-human connections, inter-human communications, plus intra-human communications are captured by the many styles and genres of literature. They also capture and analyze the Zeitgeist of this or that time, the problems of this or that society, at this or that time. The comparison of books and writings across continents and time epochs, with the myriad of content constellations applied, makes what is needed when looking for the sharpest, most sensible and open minds. Science needs to learn from literature, as literature is a reflection of all human actions and human thoughts, it is mirror and an amplifier of critical (deep) thinking.

10 Why Religion?

Compliance of the masses and a strong control of a large followership are key motivators for anyone to setup, that is, institutionalize, a religion. Some might have had, and there might have been a great deal of, benevolent ideas and notions, benevolent goals of a religion here and there. But, in the end, many of them got caught up, sooner or later, more or less, in the business of controlling the masses and deeply profiting from that control (financially, power-wise, etc.). Foucault (1975), for example, wrote about the domination of the people by the Church in medieval times and thereafter. Nietzsche saw the end coming for that system of believe and control, and hence predicted then much worse things to be happening (fascism, dictatorships, etc.), as the believe system was lost that formerly gave people hope, shape, meaning and goals. Nietzsche saw *the dangers of God (as an idea) having died or was dying*, for the most part that is (Nietzsche, 2008: 67–69).

11 Staircase to Heaven, Why?

We imagine a better life, after life? When we were young, we did. A better life in heaven. Life on earth was but a preparation, and a test, for us to reach the staircase to or the gate of heaven. So, we were told. That is what we were brought up with. A master plan. A life-long goal and always a constant warning, always accounting for a reckoning to come. No gods walk on the clouds, the heavens, and we never will either. The idea of heaven is as much a purposeful idea by humans to control humans as is the idea of hell.

Heaven and hell. A double threat, one not getting up there, one being sent down there. The rich, the aristocracies of the past, they have never cared, they were above all rules and all threats, in case they wanted to go to heaven, and not to hell, they gave money, to the Church, as a bribe, and it was fully sanctioned by the Church. Religion was for making money, then. Money went its way, and so were lots of endless powers, and the strongest of all power relations. Religion became and was for thousands of years a political and an economic entity rather than a philosophical or spiritual support system, which it was set out to be, from the onset, and how it was sold to the people, in eternity (so they hoped).

12 Our Surrounding

For Luhmann's theory, the environment, the surrounding is of utmost importance to any social entity, and any person. Where we are, that makes us. The geography of human life-propelling forces is easily to be read from national and international statistics, of mortality, corruption, incarceration, crime, short life-expectancies and much more. Where we are born we cannot control, where we live, for the very most part, for the very most of us (to varying degrees), we can control.

Choosing a great environment, and making a better environment for us, for our children, families and communities must come first, above all, as this is controllable, and it is an ethical way to control our lives, by improving our environments, hospitals, schools, public safety, and so on.

13 Daily Necessities

Hanh (1993) lectured about the need to go on with one's daily needs, daily chores and daily duties like cutting vegetables, cooking and cleaning plates. It is ok to go to the toilet when trying to achieve enlightenment, for example, by way of practicing walking meditation or other forms of meditation. It is ok, it really is. The moment in front of us, around us, is the most important in life. If we cherish our breath, our gentle steps, the earth underneath and the trees and plants besides us, we are walking on the path of mindfulness. And that is simply great, greatly simple.

14 Nature's Beauty

We need beauty, life is not easy, and beauty makes us happy. The scent of grass after mowing the lawn, it is beautiful. The earth in the woods after rainfall is beautiful. Love and the taste of vegetables, all beautiful. Children smiling, people singing and dancing. This is all plan of nature, this is all treasure, and meaning enough for us to carry on, many eons to come.

15 It Is Unethical to Look into People, It Is Dehumanizing

Hopefully the technology is not applied, to look into all people's, or into people's, thoughts and feelings. Facial micro expressions, our body posture, our social media accounts, our financial data and other online and personal data are already giving away most of our inner secrets (our treasured memories, ideas and personal characteristics, our reasons, fears and wants). To look into every one of us, into most or many of us, that will be a step backwards in the evolution of humankind (cf Kropotkin, 1902). We would become objects, enslaved and prisoners; and we would cease to be subjects, free actors and free thinkers.

This needs to be avoided. There need to be safeguards, constitutional safeguards, new policy safeguards and a new public and global awareness.

16 Be My Light, Light Me the Way

There is a great deal of people out there who are looking for answers. Some find them in random ways, some find them by running into orchestrated mass providers of answers, these include churches, sects and cults.

Some find their answers naturally in life, just by creating a habit of smiling of being content, and/or childlike and innocent. This is what Hesse's Siddhartha (1922) was looking for and what he found in the end.

Some ways of supposedly finding truth and happiness involve sacrifice, as it is mandated by some religions more than others. But this does not have to be. Anger, pride and blindness need to be shed, but this is no sacrifice, but a relief and a blessing if those are gone.

The light is always there. The eyes are the problem; the mind that is closed, twisted and confused, that is the problem. One cannot and should not ask others about the light, but simply turn to the light—a cheerful early morning can be the light, a fresh sprout in your garden can be the light, so many things that are there already can be your light.

One just needs to open the eyes and mind, breathe calmly and relax, not think too much, not think at all, just enjoy, just be. This is what Buddhism and almost all of the Bhagavad Gita (on which Buddhism is based, this is clear, if one just reads through it) teaches us; for many others, life itself has taught them (cf Swami Prabhupada, 1968).

Another way of looking at it, the problems that hinder us from happiness and enlightenment are our thoughts, our thought formations (mental formations), our culture and our social rules, the hustle of everyday life, of everyday capitalism.

These are all *entanglements*, things that pull and push our thoughts and feelings, our every action; the forces that contain us, that keep us down, that push us around. All of these, this multitude of causes, are responsible for our limited ways, for our currently chosen or, more precisely, 'stealthily' (in itself, not for itself) forced-onto ways of and in life.

17 People Themselves Have to Decide Who They Want to Be

Care for others is different, telling people what to do and finding and giving 'good'—seemingly unrefusable (as it is made to/forced to believe and/or accept)—reasons and an orchestrated reasoning to do so, that is not helping people, that is robbing them of their choices, of their right to self-determination, their right to be and choose to be who they are who they want to be (cf Bad habits in reasoning, Nietzsche, 2014: 31).

It is not the job of any government to tell people who to be and who they should be. Any government has so many other, real jobs to do, *public safety* (just public safety, and no blown-up or twisted agenda here), *social problems* (real social problems, not 'argued'/'reasoned' social problems that actually are no or no big problems), *emergencies* and *avoiding risks of emergencies* (real compelling emergencies, not 'made-up'/'argued'/'reasoned' so-called emergencies, which are no emergencies). These are the actual focuses of any proper government, and government administration or agency.

18 The Other Side

Modern-day people, and civilizations, are proud, so proud that they cannot see over to the other side. They are narcissistic to a point of calling, or trying as much as possible to call, everything else barbaric, or undeveloped or underdeveloped, or failed. And we do not see the barbarism in much of the things that is going on today, amongst us, in between these and those of us, next to us.

So, then, how can, how could we see beyond the walls of time? How could we imagine to live and share the world of the ancient Greeks and Romans, the Incas, the Azteks, and ancient kingdoms and empires in Africa?

The walls in between us and around us today are blinding, at best they blending myth with reality, blinding out the bad and emphasizing the positive. We need to work on seeing these walls of today, and those that separate us from the past.

Comparison is key, switching on and off the lights is key. This is the Kantian-inspired method of looking at the evidence (Kant, 1781), the empirical world (the data), then looking up to theory (build one, or ideas that lead to one), and then down again to the empirical, and then up again, and so forth, as one side is never enough to see what is going on. Patience is key, too, and so is an open and critical—deep-thinking, all-around-thinking—mind; a mindful mind.

19 No Wood, Only Trees

Entanglements—compelling and obstructing inter-complexities (forces that are pulling us, pushing us)—are there right in front of us, all around us, but we cannot see them. We cannot grasp them. What is not named cannot be understood. What is not sought cannot be found. What is not explained cannot be rallied against. The forces of a myriad of entanglements all play together, they play a harmonious symphony, and we do not know it. The results we know, we see.

Poverty, unemployment, imprisonment, traffic police punishing people and making money from punishing people, children being scolded by their parents and suppressed by homework and exams, the sick and the elder citizens being shamed and blamed, we are always shamed and blamed (cf e.g. Ryan, 1976; Maine, 2017). We eat meat, we are overweight, we drink beer and wine (i.e. 'alcohol', what a 'bad' word), we smoke, we like to have sex, we like to binge-watch soap operas and TV-series on our television or our phone, we seek pleasure, so we are shamed and blamed, by religion, by the official media and the social media, in school textbooks, in bibles (and scripts like that), as well as punished by government taxation and regulation.

We have to excuse our behavior, sacrifice our needs, we have to give in and be quiet, we have to follow and swallow what they put in front of us (goals, directives, rulings, regulations). We are forced to vow, to show remorse, plead guilty (cf Foucault, 1975, 2014).

We have to live a dehumanized life, in degrees dehumanized, sometimes more, sometimes less, in some aspects more and in some less. We are forced to follow wrong idols; maximizing our consumption of goods and services and the expenditure of our financial resources; maximizing the use and abuse of the earth's resources; maximizing the exploitation of our mental and physical resources (children in disciplining and torturing schools, parents working overtime and double jobs, exhausting jobs). We are denied decency, humanity and (our, the common people's) happiness.

And in all of this, entanglements that run through time, and across continents, have their hands on this, they are pulling the strings, rigging the game, rigging life.

20 Sumak Kawsay/Vivir Bien/Buen Vivir

The Andes are home to an exceptionally important concept, *sumak kawsay* (in Kichwa/Quichua language). Internationally it is known under the Spanish names, *vivir bien* (in Bolivia) and *buen vivir* (in Ecuador). Its key concept of 'community life' (*la vida comunitaria*) can be found in indigenous societies all the way from Alaska to Patagonia (cf Huanacuni Mamani, 2010). While they still had their own world, without interference and extinction tendencies brought by Spanish invaders, the native people of the Americas, have created this wonderful welfare- and ecological-oriented set of ideas, philosophy, and worldview.

Sumak kawsay is focusing on more than one dimension in development. It advances a parallel, and equally emphasized, focus on material (economic), social and spiritual well-being (satisfaction) for all members of the community, not at the expense of others or other communities, and not to the detriment of the natural environment (no excuses, no exceptions there).

That this is possible is to a very large extent the result of *distance* to, and hence *isolation/protection* from, negative European Colonial influence and domination (cf Adam Smith, 1776). The *non-entanglement* of the Andes has protected sumak kawsay (vivir bien/buen vivir), luckily due to the high mountains of the Andes that provided physical and hence cultural protection. The concept of sumak kawsay has survived and, in recent times, it has been experiencing a renaissance, growing international appraisal and acceptance.

21 Letting Ourself Become Abducted

It's a game of sacrifice. Built on soft and smothering lies. You have to submit, your intellect, and let freedom fly. The peers are waiting, and they play their game. The wearier ones have more to lose, and choose to push all the more.

Let yourself be pushed, or you perish. Be kind and cunning, let them deflect on you with their tricks of world domination. Hush, hush. Who said anything about freedom, innovation and remorse?

We are in—ok. We dance with you, the forces of the ones that hold the steering wheel and all the powers to let us on and be on this or that ship. We need to set sail, and go to sea, and many days and years later, then they will see.

22 Temptation

A clearly self-propelling force, temptation breeds temptation, and it is an addiction, a progressive force, also a self-preservation force; it is sticky, it likes to stick around, it is enduring (an enduring force).

Our daily temptations are best taken care of with regular routines, thought routines, behavioral routines and familiar, stable environments (that accompany, i.e. match, our routines). These daily regiments, thought systems/regiments and other, behavioral regiments, are helping us to protect our good temptations (exercise, taking a nap, meditate, oatmeal, 80% raspberry marmalade, work, those things), and shut out and protect us against unwanted—actual, factual bad, i.e. unhealthy, and perceived/trained/told bad—temptations, e.g. accepted/perceived to be immoral things.

Our own regiments—*reoccurring behavior and thoughts, thoughtways and thought systems*—are there to protect or prevent/protect from (respectively) wandering-out thoughts and wandering-out behavior. Too much regiment is not good either, it kills innovation; too less can be detrimental, e.g. to our mental health, its stability.

23 Insecurity

People everywhere are looking for instructions and patterns, to find them and to follow them. We are, the most of us, to a high degree, with regard to many things, expecting (and liking; and most do not like to admit it) to be told what to do.

This is a secret, and open secret rather than a hidden one. That is one of the key differences of humans that makes us human, as compared to apes or cats. We are willing to learn, even to accept knowingly faulty knowledge (cf Horner and Whiten, 2005: as well as e.g. Povinelli, 2000; Seed et al., 2012; Manrique and Walker, 2023). We accept to go to school and sit there for many hours almost every day, for so many years. Other animals (other than human animals) are *not* keen to accept—(and hence reject)—instructions.

Apes do not like to be told what to do, and if, they like to not take it for granted, that is they like to reinvent the wheel, and that almost every time when someone is teaching/showing them about this or that (cf e.g. Horner & Whiten, 2005).

Humans are prone to learn anything, including ill-advised, out-of-date and unpractical things, i.e. all things. The high development of language, plus the storage of language (books, etc.), in addition to the nature of humans to accept to be told things and willing to repeat those things (in most cases, for the most part, *no matter what*) has led to our development of cultures, large and complex societies and their economies, and of course technology, with all their ('crazy'-looking) latest and highly progressive achievements and devastating consequences.

The modern scientist, a human too, likes to search for and ask for patterns. The scientist likes to follow manuals of methodology, copy this or that. Clear definitions are asked, and many times they remain withheld. Often clear definitions are presumptuous, tainted or random—due to our perceptions, ideas, preferences and expectations; and due the past-laden, context-laden/culture-laden meanings of the words, any words, we use. In addition, the modern scientist likes to stick to familiar, easy-consumable (*leicht konsumierbare*) theories. That explains the high popularity of many theories (and models), and the washed-ashore phenomenon of many others that have been ever found, then neglected and/or forgotten. Easy 1-2-3-style classifications, simple models explained by simple boxes and pyramids (and often they *are* helpful) are the food for the masses, of scientists. The easier a theory, the ever greater are its progressively increasing potentials for historical and global acclaim, this is a general tendency.

The existence and prominence of Kuhnian *normal science* (1970a, 1970b, 1977, 1978)—i.e. outdated, averaged-out, plus spun-out-of-proportion/running-wild (ausartende), purpose-estranged (zweckentfremdete) and solely self-promoting, twisted and perverted at the same time, theory making and theory testing—is related to this.

On the one hand, insecurity causes Kuhnian *normal science*, and then again Kuhnian *normal science* of course causes, strengthens and thus helps to directly, and indirectly, multiply insecurity all along.

The average (i.e. middle-along-the-road) scientist is a human, scared and disappointed, especially if they know clearly where they want to be, and where they are not and/or cannot go, ever. This leads to insecurity. However, insecurity, also has its feats, it is helping to perpetuate theories and practices, it leads to enhanced, commonly agreed practices. The downside includes a relative lack of innovation, a relative lack of excellence as a common event (in the majority of all single scientific works). And, another key consequence of the rather general insecurity of scientists (i.e. those that are insecure) is that young scientists get shot down by older scientists due to their acquired powers (as frequent reviewers and hence core gatekeepers), especially and almost all the time, when new ideas, new methods and new theories are introduced or supported (cf "Arrogance has always been justly called the 'vice of the sage'" as their works have become their 'ego', Nietzsche, 2014: 280).

24 Hierarchies

The human is a rank-oriented animal. Groups of higher-ranking humans team up to suppress the others, and where there are dictators, they suppress all others. We are no exceptions in this regard, in the animal kingdom.

Almost everyone, given the chance, likes to be a dictator, or over time tends to be like a dictator (cf Milgram, 1963, 1974). Leading groups of people like to dictate all things to all others. Any group put in charge will, given time, sooner or later, behave like this.

Any common human-made institution or set of institution will also behave like this, if given the chance, and time. The welfare state is no exception. It ever more finds ways to tell all people what to do; except the very rich and the very powerful. The power elites (as their name indicates) find ways, and always have, and always will, to influence, change to their advantage and prevent or alter for them unfavorable conditions—with regard to financing mechanisms, taxations, punishments, benefits, laws, rulings, rules, regulations, institutions, policies and programs; any of them, all of them.

The welfare state of today is a system that controls, and not one that liberates as early proponents and followers had in mind; the same is true with most churches and religions, or political liberation movements.

Chapter 10
Back to Quantum Methodology

1 The Wonderful Prospects of Quantum-Inspired Methodology

New things are possible. As with quantum technology, so it is also with social science. New inspirations are leading, can be leading, to new shores and new heights. The old gets questioned, and the new gets a chance to proof itself. A new showoff between the established views, theories and methods, and the now slowly and continuously, or rather fast or fast mushrooming ones.

Randomness, parallel theory development, parallel theory-cum-method development, and the further development of *abductive methods* are some of the perhaps most prominent and most promising features of quantum-inspired methodology, outside the technical and natural sciences.

Behavioral, life and social sciences are, already for some time now, particularly interested in this new sort of venture, and indeed very exiting direction of setting out on new research programs, that may stretch over years and decades; with many new breakthroughs and developments up the sleeve.

2 Abduction, the Cornerstone of Scientific Knowledge Creation

Abduction, *as a methodology*, is integrating both directions, the inductive and deductive research actions (in an alternating manner): by way of switching/alternation in between *hypothesis-setting/-improving* (empirically-and-theoretically-informed and -based, provisional, i.e. non-final, decision-making), and then particularly *deductive action* (theory-choice, theory-sketching, theory-building, theory-finetuning, theory-improving, theory-checking) and *inductive action* (analyzing empirical evidence and

C. Aspalter, *Human Entanglement Theory*,
https://doi.org/10.1007/978-981-97-5161-7_10

data, trying with them to substantiate or overthrow a theory, i.e. testing/trying to negate a theory).

With every research venture, oneself needs to find and determine the exact steps of actions taken. There are different interpretations and applications of abductive methodology; of course also varying from one discipline to another, and also within one and the same.

Kant (1781) did kill the method of *pure* reasoning *and* the method of *pure* empiricism (inductivism). Popper (2009) then (again) killed *pure* induction methods as scientific methods (i.e. methods of knowledge creation).

Coming up with sentences describing the data and evidence is nothing but an interpretation of the data, of empirical findings. One can never, so Popper, make a universal law or theory out of data or empirical findings alone. The proclaimed universality of any theory or inference to theory from any inductive-method is non-existent, hence scientific nature of such theory or inference is not there, so Popper.

In general, only *the method applied* can guarantee any possible or actual scientific nature of sorts. Deduction from theory, and hypothesis, and checking against data/empirical evidence, so Popper, is the only way to arrive at any scientific output and conclusions, new scientific findings.

Pure reason (as Kant made it clear) are not empirical theories, and do not belong to empirical science; they are part of metaphysics and philosophy only, and hence not science.

Apart from this, König (2022) elaborated on and explained to the point the unfinished business of Kant's theory of methodology, in terms with theory building, and science creation. Kant's important contribution was to deny pure reason the status of science, and to deny as well pure empiricism also the status of science. Popper focused the latter finding, but also helped to specify (fine-tune) the first one. Kant and Popper also point out (and the author fully agrees) the problem of (final/deterministic) demarcations between different steps and concepts within the scientific method that are troublesome, if not impossible to make.

Popper with all his elaborations left out something, and biggest and most indispensable of all: the presence and importance, and the particularities and functioning of *abductive methodology* (cf also Pierce, 1878a, 1878b, 1893; Schurz, 2008; as well as the unfinished business/task of Kant, see König, 2022).

All knowledge creation in all of science is in fact done by abduction methods. We do not see it, we do not talk about it, and we have not understood it yet. The hypothesis-setting and negation-based testing of these hypotheses and theories is mostly done by students, in thesis writing, but also by many short essay type research works where the topic and research object and research questions under scrutiny *are always changing* from one paper to another. The rest of scientific works are conducted, in fact, differently. The usual model is research programs, a *series* of research works on the same theory and/or following the same research questions; either by the same author and many different studies by different authors.

Most individual researchers work in this way, as well as most teams of researchers, and interest groups among researchers of the same discipline (or more disciplines).

Every piece and study of research, within such research programs or research groups, is merely a step in "*one*" research *process*. And in this research process, the method applied, as a whole, is the *abductive method*. As past results are flowing in and form the basis of new hypothesis, new supported theories that, with every new research work, are being tested; that is, their predictions and findings are being compared to empirical evidence, *again and again*. Theories are formed, built, substantiated or refuted/rejected, along the way. We are talking about chains of research works, where each research work is contingent/reliant—more or less, and in general—on former research works.

Here it needs to note, *quantum-inspired methods* in social science are *adding to the variety and capabilities* of general abductive methods, and hence abductive methodology as a whole.

In particular here, for example, it can be noted that (1) the *switching* between theory and empirical evidence is increased and even compressed into and within one research study; also (2) theory building, and particular method building (i.e. discipline-specific methods, plus research-purpose-specific methods) are now also encouraged (allowed) or enabled to proceed in parallel.

Moreover, (3) the '*universality*' of research studies (i.e. the *exact* same use of definitions, concepts and their interpretations *all the way*), within any research program or group, is additionally emphasized, which was lacking or formed the Achilles heel of many research discussions and theory development attempts in the past, that also involved various forms of research triangulation (researcher triangulation, theory triangulation, method triangulation, case study triangulation, data triangulation, etc.).

3 On Universality of Infinite Vectors

Within the arena of the human entanglement theory, the forces (entanglements/vector forces) that are looked for, and looked at, under the analytical microscope are to be matching in terms of their *least common denominators*.

The theory of human entanglements, in a post-Luhmannian tradition (where I extended the theorizing of communications to also include all intra-personal communications and human-cum-machine communications), uses pieces of communication, i.e. *words* (plus feelings and art, that yet again can be expressed/represented in words) *as the smallest—and commonly shared smallest—unit of observations.*

Words are the smallest common elements of thought systems, emotional systems, as well as social systems, such as sentences, theater plays, conversations, poems, policies and laws, etc. (all of which are systems of words).

It is necessary to ensure compatibility of observations all the way, this is what is understood as 'universality' in the context of *high-dimensional, infinite* vector analysis of human entanglements *of different kinds*.

4 On Universality in an Infinite Universe (of Observations/ Events/Cases)

Infinite is the number of observations (events occurring or entanglements), as well as the complexity of their inter-connectivity (in our case, entanglements of entanglements). How to operate in an infinite universe?

This we can learn from quantum computing, the principles and strategies applied rather than the exact solutions (but also from the solutions).

5 Communication: The Least Common Denominator

Mathematics has solved the operational conundrum of quantum computing. And, in quantum social research, too, *logic* is needed to make sure quantum approaches and quantum methodologies are working in the world of science, and philosophy.

In our case of looking at a myriad of enabling, compelling, obstructing and preventing human inter-complexities (now a.k.a 'entanglements'), the atoms of human action and human communications are always words, or things that can be expressed or conveyed in words. The *method of the least common denominator* is used in quantum computing. It also needs to be used in all sorts of social research that use quantum-inspired methods and the therefrom developed theories.

6 On Parallelity: Its Necessity and Advantages

Working in tandem, working on more than one research work, or research aim and field, this is not an alien concept for the very most of us, researchers in current times. Quantum computing does the same, but is has taken this to its extremes. That is how, in essence, quantum computing is so revolutionary and has come to exert—from a societal point of view—an all-changing effect (literally almost *an all-changing effect!*; cf e.g. Aspalter, 2023c).

For social science research, in the area of quantum applications, too, the issue of parallelity may (as this book demonstrates) serve as a vital guiding principle in setting up and conducting quantum-inspired research in the any social science and humanities discipline.

For the positive effects of parallelity, this will depend on the particular application of quantum research (theory, method, etc.) and purpose and design of the study undertaken. For us, here, in this particular study and setting, parallelity was most essential, and truly indispensable. It is thus imaginable that in the majority of future quantum-inspired/quantum-based research studies, this is also the case. Conversely, it will be perhaps not common, and very uncommon, not to have parallelity as the key-enabling ingredient. Hence, there may be a few exceptions, but in general, for the bulk of research studies yet to come, *parallelity can be expected to be a constituting and completely indispensable characteristic of quantum-like research in the humanities and social sciences.*

7 Dialectic and Its Different Meanings

In Ancient Greece, there were also many interpretations of what dialectic is, as there have been since Immanuel Kant, Georg Hegel, Karl Marx and Friedrich Engels, and beyond.

The most here useful, and here used, interpretation and definition of *dialectic* is the one where the illusions of this or that concept (categories, ideas, methods) are being overcome, they are shown to be misconceived, by switching from one concept to its opposite concept, and then back again, and that in a repeating manner.

> One can imagine, if a room has such a bright light that one cannot see anything, one can turn off the light, and then yet see nothing, as all is pitch black. That is also true the other way.
>
> But, if one keeps switching on the light and off the light (and if one does not mind the pain of looking out in the room while doing so) for a long enough time (and if one has a light-speed capable sensory system!), one then can possibly see what is going on, has been gone on, all the time, in that room.

Immanuel Kant came closest to this, in and with his thought concepts and philosophy. However, he understood the term dialectic in common people's terms, and not anything close to what most academics are using it nowadays, or in the old days, in Ancient Greece. Kant, of course, did not name the method, he was about to be finding, and fell short of just finding it (cf also König, 2022).

Hegel's, Marx and Engel's concepts are not useful in our endeavor, and diverted the brilliant line of thought started by Kant.

> For methodology, this definition of dialectic that is used in our book here means that both (pure) inductivism and (pure) deductivism are wrong in general. They may lead to good results here and there, if the researcher is lucky. But science is not conducted by space-travelling tourists (aliens) who come to earth and have no prior experience, do not speak local languages, and are not at all acquainted (and tainted) by our cultures, i.e. cultural bondages (our cultural 'baggage'), and have not read or heard about prior research outputs on earth, and each time they have finished one research work they will leave earth at once, in different directions, and not talk to the other space-travelling aliens, who will also come and conduct another, either purely inductive (and singular!) or purely deductive (and singular!) study or experiment and so forth.

As long people use words and sentences, use languages that have long histories of usages, interpretations, connotations, evaluations, taboos, fears and destain, enthusiasm and liking, included in each sentence and most of the words (and all keywords) used, induction will always also be, inherently, deductive (Popper, 2012, referred to this as somehow, and weirdly, "quasi-inductive"; the question is here *does a quasi-poison exist?, a quasi-solution?*).

Similarly, pure deduction, too, is spoiled (tainted) by world experiences, history (our story) and its results (outcome, empirical evidence), by experiences of the researchers themselves, and especially also the other researchers' findings and theories that also have life experiences. By either way of life experiences of oneself (and that of all other people on earth today, and in the past) or scientific empirical findings of other research studies and researchers in the past, deduction is always also induction, to a smaller or larger extent.

Nothing is pure, nothing is clearly possible to be separate from its other extreme. Induction will always be 'spicked' with lots of 'fat' (rich, complex and stealthy) meanings, in each and every sentence of its interpretations and evaluations of their data, as long as language is being used, as long as the eyes and minds of the researchers (here on earth) are being used. The same is true with deduction, data and experiences will permeate, and riddle, every so deductive process, no matter how purely it is perceived, no matter how strongly it is intended to be pure.

The consequence is that one can only find, with no fallibility attached to it, that all of science is using both! Both are wrong, when defined and understood as extreme, pure concepts. They are both wrong, when they are used as mutually exclusive *and 'entirely' and 'forever' separated concepts.*

Both are right, as long as it is understood that the one builds on the other (both build on each other)! No one needs to be shamed, no one needs to be scolded, for either being an inductivist or a deductivist, as being either one, always includes the other, to a lesser or greater degree, and direct and indirect degree. This is so by way of past experiences, past findings, the path dependencies of ideas and concepts, past and current language usage, as well as the path dependencies of any theory/idea/ideology/concept/notion, common logic, and all scientific thought systems and practice systems.

In a nutshell, *all of science is but abduction*, as the wheel of science is indeed a wheel that is turning, being spun also by all previous researchers and their studies, previous theories and ideologies, and previous methodologies used and their findings reported (cf Pierce, 1878a, 1878b, 1893; William, 1970, 1981; Dewey, 1910, 1922).

This is true, except for the (hypothetical) occasional alien scientist, on a short inter-stellar stopover, who is, and has been always, isolated completely from earth, its researchers, their findings, their common logic that they use, as well as their language and culture (habits and traditions). But then again, this travelling scientist will be subject to alien concepts, alien logic, alien common sense, all of which are also tainted (again) by their past, their language, their logic and traditions. So that alien scientist must have been a baby at the time of the departure from his/her/its home planet, and not instructed or taught by others (nothing, not their language, not their culture or logic) on the way to earth.

With this thought experiment, hence, one has to come to the conclusion that no one is ever conducting science without being subject to the method of abduction, everyone is doing it, and will be doing it, and that in all of the universe (and in others, if there were others)!

Chapter 11
Back to Human Entanglement Theory

1 Entanglements and Their Vector Forces

To be clear, human entanglements are not to a hundred percent deterministic, and not to a hundred percent determinable (as they cannot be caught in a timeless manner, and/or a finite or perfectly complete quantifiable manner). First, there is always room for freedom, for butterfly effects to roam the world of human interactions. Second, the either obstructing and/or compelling forces of entanglements *operate cumulatively and with likelihoods*. They are like vector forces, in complex vector spaces.

For future studies, in the social and human sciences, the same technique as in quantum computing can be used, e.g. by transforming complex vector spaces into geometrical *Hilbert spaces*—that facilitate mathematical calculations, of otherwise infinite events/results/vector forces. Hilbert spaces are used in quantum computing, as a main means to operate with and calculate data (cf De, 2020).

The *precise* vector values of groups of relevant vectors are unknown. That is, they are (*all* the relevant ones) *not* known to the exact point—from a universal standpoint, from the perspective of universality. One needs to ask each time, which ones are most relevant, to what degree, to the first degree, to the second etc. degree, and which not, and why not?

One can measure them *here and there* (e.g. punishments in form of payments, costs, etc.), and one by one, at one given time, but their aggregate, fully aggregate (i.e. universal, total) values are to be determined again, each time, all the way, once time has moved on (to a lesser or larger degree). Things change, and all things are related to all things—that is why. These vector values *change*! So, human entanglements are in general dynamic, and not static in time. They change, they are alive, so to speak!

One cannot measure *everything* (everything relevant) to the exact point, according to the *Heisenberg's uncertainty principle, or the principle of blur relationships (of which there are two), i.e. Unschärferelationen* (Heisenberg, 1927a, 1927b; SF, 2023). Schrödinger (1935a) called it the inaccuracy relationship, *Ungenauigkeitsbeziehung*,

C. Aspalter, *Human Entanglement Theory*,
https://doi.org/10.1007/978-981-97-5161-7_11

which explains all-existing tolerance and variation margins, i.e. *Toleranz- und Variationsbreiten*. In addition, the *observer effect*, which is a different principle (and often confused with the uncertainty principle), says measuring itself always changes the values one wants to measure (cf Baclawski, 2018; Zimmermann Jones, 2018).

In the macroscopic world, the observer effect means, when one asks too many questions, for example, the one being questioned starts to learn, and starts to give more and more socially desirable answers, politically correct answers, more and more self-reflective answers, and the like (cf Baclawski, 2018).

Having said that, it is *quantum computing* that pushes ever faster than before the boundaries of what is possible and what not. The *combinatorics capabilities*, the capabilities to predict ever better and ever longer into the future, of quantum computers are speeding up progressively.

Hence, to some degree, and to a fast-growing degree, maybe not so much humans, but more human-made *quantum calculators* (quantum computers) are capable to look—a bit farther—into the future, and to determine future outcomes based on past and current human communications and actions (cf e.g. Greenbaum & Gerstein, 2023). Nietzsche has already looked at the problematic possibilities of determining future outcomes, including the general delusion of a (strong or any) free will:

> In looking at a waterfall we imagine that there is freedom of will and fancy in the countless turnings, twistings, and breakings of the waves; but everything is compulsory, every movement can be mathematically calculated. So it is also with human actions; one would have to be able to calculate every single action beforehand if one were all-knowing; equally so all progress of knowledge, even error, all malice. The one who acts certainly labours under the illusion of voluntariness; if the world's wheel were to stand still for a moment and an all-knowing, calculating reason were there to make use of this pause, it could foretell the future of every creature to the remotest times, and mark out every track upon that wheel would continue to roll. (At the waterfall, Nietzsche, 2014: 69–70).

2 A Theory of Human Communication and Human Action, and Its Implications for Personal Freedom

This theory, as set forth in this book, has all of human communications as its research object. It observes, investigates and hence encapsulates the entire history (ourstory) of human communications, including inter-personal and intra-personal communications, as well as machine-based—machine-enabled and machine-generated—forms/modes of communications (as humans have made and programed the machine in the first place, at least the first machines that started to communicate with, program and/or make other machines).

Human communications are closely tied to human actions, but they are not identical. A great deal of human actions are—function as—human communications, but not all. Depending on human context, or human-made/machine-made context (e.g. cameras observing and computers analyzing, etc.), human action can be understood and received, and hence seen, as human communications. Like e.g. when one eats a banana in a board meeting, or leans back relaxed in the chair, in the very same or any other meeting. So, it depends really. The theory of ours here is also including human actions that function as or are understood (act) as human communications. That could e.g. include blowing a kiss, dressing up nicely for an interview or a date, choosing the right color tie and shirt for this or that lecture, and so forth.

The primary concern and interest of the *Theory of Human Entanglement* is with people in general, i.e. when looking at millions or thousands, or hundreds of people at a time. Hence, the degrees of rather limited freedom predicted are that of average, median people in and conditioned by very same situations, events and entanglements (language, culture, the past, laws, practices, economic circumstances, social conditions, and so forth).

Single people can surely break out of Plato's cave (2011: 514a–521d). They can work themselves out of the cave, e.g. if they can master the art of prison break, e.g. to pick a lock (figuratively speaking). Turning away from the crowd, from the information feed of the past and the information feed of all others and especially the ones that are in control (or in a dominant position; influencing, pushing, and nudging all others) is extremely difficult.

> Information is 'reality.' Our *perceived* 'reality' is limited and created at the same time through information that we receive, observe/invent/create and process (cf Zeilinger, 2005, 2010; Gröblacher et al., 2007).
>
> That explains the forces of *Plato's cave*, the forces that create and administer it. These forces are *human entanglements*, they are based on and created by words, i.e. information.

Even more difficult is realizing and minimizing the effects of the deep-into-the-past-reaching, stealthily-hidden and deep-into-the-consciousness-and-subconsciousness-reaching language tunnels. And then, there are the personal habitual, social and cultural strings, the last dimension of influence and in larger part predetermination that one needs to leave behind, bit by bit, one step at a time.

Hence, breaking away, i.e. slowly breaking out, is very possible, but not for the many, just for the few. Plato was keenly aware of this. Krishna and Buddha they also knew, and tried to bring relief. Plato was more positive and optimistic; and the other two were more nihilistic (cf Nietzsche's thoughts on Buddhism, Nietzsche, 2007: 118, 146–149).

It comes with a steep prize, of loneliness and isolation. Plato also mentioned pity, anger, and violence. So, one shall be diplomatic and at least accommodating (to the 'game', to this or that) in terms of trying to stay alive (look what happened to Socrates, Plato's teacher, they killed him literally for not falling in line, for not worshipping religion in their ways, for having different opinions and for teaching them to others, and this was in the name of *democracy*, applying the rule of majority vote, Socrates was voted to be executed, by a majority, amongst 501 people, wow, what a story, what a rich and deep-and-far-reaching story) (cf esp. Plato, 2009).

3 The Knowing and Calculating of What's to Come

It was and always will be the greatest driving force of humanity, the greatest of all (earthly) powers come with it, to be able to predict what is to come. We started out tens of thousands of years ago to observe and measure the wanderings of the sun and the stars, in order to predict the future. The times of sowing and harvesting were particularly important. Observing sunrise and sundown points and marking them were key.

We are at a point in the story of humanity where we again make giant leaps forward in application of knowledge (science), and here again quantum computing technology. Things will never be the same, fortunately for the few, and unfortunately for the many.

4 Utopia, Dystopia

In general, socialism was for long, and is still for many, the solution, the utopia, a (*supposed to be*) 'perfect' version of the world that needs to be achieved, no matter what. The failure of socialism as a full-fledged system of societal and economic order to provide evidence of ever achieving such a goal speaks for itself.

One person's dream, dogma and perfect solution, in general is some other persons' dystopia. Too many things are entangled to one another. Solving one thing has implications on all other things.

The economy needs to keep running, people need to keep working. China has learned all that, and thus became a most 'positive' example of socialism, compared to other forms and practices of socialism, due to its success of achieving great economic success over already a very long period of time.

Other older forms of socialism died because of the lack of the use of proper cost accounting in particular, and hence the lack of a workable/functioning cost-containment and cost-reduction system; this was noticed also by e.g. Kim Il-Sung, upon recommendation from the Soviets (cf Woo, 2022: 140–141).

Capitalism per se is not better, is not the solution to everything, but also the source of many problems. Socialism does not only bring solutions, but also many problems, some are solvable and some not.

A mix? Most countries in the world today are on the path of mixing the one with the other (Germany, Sweden, China, all of them). How the mix, this is the key to success. The core problems of socialism *must* be avoided, as are the core problems of capitalism. The particular solutions to problems and the particular ways to prevent and remedy any old and new problems are the key.

> Our glasses of language, of culture, of politics, of ideas and concepts that are spiked with hatred and bigotry are always on when we speak, and think. We are barely able to objectively discuss socialism versus capitalism, without falling into a trench war of thoughts (ideas) and thought systems (ideologies), of misrepresented and misunderstood versions of the past and the present.

Cost accountants and policy consultants are needed, *not only* for all socialist countries, socialist governments and parties, *but also* for all capitalist countries, capitalist governments and parties, all the way.

With socialism, the economic problems are not solvable with socialist solutions of dictatorship and command economy. With capitalism, social problems are not solvable with market, competition and choices, as these are the causal forces behind those social problems; disguising them (due to political and ideological reasons) as solutions does not help, but worsens the situation (cf Piketty, 2014, 2020; as well as Galbraith, 1954, 1958, 1992, 1996).

The ways the welfare state has done things in the past are second best, at best. In most countries (in Europe and in most former European Colonies around the world), some of the main solutions proposed are the main reasons for the weakness of these welfare states, their societies *and* economies. Other solutions seldom applied (except for in Singapore and elsewhere, all scattered across the globe) are the solutions that still await application worldwide, yet they solve both social and economic problems, or solve social problems and avoid additional economic and/or financial problems (Aspalter, 2023a; Aspalter & Teguh-Pribadi, 2017).

5 Improvements

Any defeat is separated from success by yet another attempt that turns defeat into success. Generally, that is how it is. One can always do something, it is never too late or completely hopeless/useless. In world politics, in global economics, in sustainable development as much as in personal finance and health/lifestyle interventions, family affairs, work-wise and job-market-wise. We need to keep trying. And, we need to keep learning. These two go hand in hand. Resignation (acceptance or refuge into nihilism) is not the answer; a positive attitude to struggle in life is.

One has to fight for one's health, work for and on one's family, and plan one's resources (time in particular, but also incoming and outgoing finances) and choose wisely (the best one can, according to one's capabilities at this or that point in time, under this or that conditions).

Learning is built on reflection, and it is also built on failures. Failures are ok, some are good, as long as we learn from them, as long as we take away a solution that avoids this problem or failure in the future, and turns an unsuccessful past into a successful future.

6 The Entanglements of Social Problems

Rationality is spiked with entanglements, or *Forms* as Plato would have called them (Plato, 1990: 126a-166c, as well as 2009, 2010, 2011). Forms are entanglements, which are the product of entanglements and yet cause, here and there, plenty of entanglements themselves. Words, the very most part of them, are heavily and stealthily tainted. They are steered and hijacked by imaginations, ideas, concepts and connotations.

Rationality likes to explain—i.e. people who use rationality like to explain—social and societal problems. Yet, we do not get reality from these explanations, but merely representations of representations of reality that are soaked and riddled with, hooked up and hijacked with, an infinite number of semantic (denotational and connotational), etymological, socio-cultural entanglements, plus current-day and past-day usages, and their path-dependencies (entanglements of entanglements)—e.g. words were used for what, to say what, to do what, by whom and how do we think about them and what else they have been doing, and that today (and not back then), etc.

Social entanglements, political entanglements, too, are having their share of distortions of meanings, of reported and actual intentions, of aimed effects and solutions. The problem of entanglements (of all sorts) is yet another *meta problem*, for society, for our cultures and civilizations, and the world (our mother earth, that we manage to destroy, and not manage to rescue).

When dealing with social and societal problems, one always also deals with lingual, cultural, legal, administrative and common sense traditions (see Gramsci, 2011a-c). The past flows in us in a hundred ways, so Nietzsche (Whither we must travel, *2014: 343–344), it really does.*

7 Diseases and Their Non-biological Causes

In health policy, what people talk and feel obliged to handle in terms of government policy, action, and recommendations for the very same is not up to speed of scientific developments ever, and one can wonder why (inertia, procrastination, and such).

The ways of general people, the experts, the scientists, the policymakers, and the rank and file administrators are channeled by group dynamics of people's ideas, goals and motivations (and non-motivations). We can understand these as *group entanglements*, entanglements that occur and like to occur in group, or *en masse* (cf also Le Bon, 1881, 1894, 1895, 1918).

Diseases can mostly be understood with current-day science, and prevented, especially modern mass diseases (i.e. non-communicable diseases). Yet, their perseverance and rampant spread across all countries and regions of the globe gives rise to nothing but worries, and the questions arise, what goes wrong here?, what is wrong with us?

We, as humanity, as societies, as well as scientific and intellectual leaders, are not evolved enough yet to recognize successfully *and* deal successfully with the problems that have been presented and elaborated to us by Plato (2011: 514a–521d), with his cave allegory.

Business as usual, interpretations and practices as usual, the Kuhnian tunnels of 'normal science' as usual (Kuhn, 1970a, 1970b), all these are fabricated, fitted, cemented and expanded over time, as entanglements like to reproduce themselves, and create spinoffs as well.

> Human (cultural, linguistic, historical/ourstorical, etc.) entanglements are living entities—ideas live, ideologies live, entanglements live.
>
> Language lives, culture lives, practices live, and we live through and with them.

Even after our passing away, when we enter into another sphere, in that sphere, we act no more, we talk no more, and yet our words and past deeds are still acting, are still received, interpreted, understood, dissected and analyzed, reworked and thrown back into public or other discourses. They then are all *living* memories, *with yet lives of their own*, so to speak.

As long as we live, in order to change the way we live, we choose to live healthily, not so healthily, sometimes healthily, or never much at all, we need to deal with *the forces of entanglements.*

How much people love red meat, love their cold beers, love white rice, white bread and so on, these are the result of cultural and personal entanglements. *White rice* in China, is understood and treasured *today* like the holy cows of India. The same is with pork, beef, or any meats in much of the world. To say, eat less of it, maybe only half, is a kind of *sacrilege.* Indeed, this amounts to nothing else than a modern-day *blasphemy*, for most people in most countries.

And these traditions have grown and spread sometimes over thousands of years, and sometimes over just a few decades (cf e.g. Rada, 1983; Qian et al., 2021). Before the 1980s, poorer people here mostly or entirely ate sweet potatoes, bananas, pumpkin, etc., instead of rice which was labor-intensive and hence expensive (cf Rada, 1983). Today, young and mid-aged people in the South of China cannot imagine not to eat *white rice* every day now (more than once).

Unhealthy white rice (which is for the very most part made out of simple refined carbohydrates, and thus identical for the most part to sugar), is made by the peeling of whole rice grains, i.e. removing all *the healthy parts* from the original rice grain. This new way of eating rice, and also the shift away from health other foods (sweet potatoes, bananas, pumpkins, etc.) have left their mark on the health of a large part of the human population, here in Asia. The problem is that people will not easily give up their holy cows, they will not easily turn to oatmeal for breakfast, and vegetables and exercise as a way of life.

In conclusion, much diseases of today, the modern mass diseases in particular, are the outcome of the making of cultural and lingual forces, historical (ourstorical) forces, personal and communal forces, that not only act as the puppeteers of our life story, of our lifeworld (community, work, school, friends, relatives, etc.), but also of our national and regional lifeways. The *puppeteers* (our entanglements) are orchestrating all of these, and formidably so. These puppeteers are our entanglements, our multi-dimensional *Fortuna* (in the sense of Nietzsche, The tragi-comedy of Regensburg, 2014: 346–347).

8 Activation and Prioritization

Most of the human DNA is seemingly not used, and from the ones that are there in general to be used; many are in a non-active state, for all of our lives, for most, or some parts. Some DNA's job is to activate other DNA, or to prioritize one over the other, and sometimes switching from one to another (cf e.g. Sanders, 2021; Wu et al., 2023).

Changing eye and hair colors are signs of and evidence for this. Take for example, as I happen to be able to speak from experience, my own eyes changed from grey-green (my passport said that when I was a child and it was true) to for the most part blue with orange/gold areas at the center; my hair too used to be super blond and now it is rather dark brown, for the most part (this happens quite often where I come from).

This however does not mean that turned-on DNA has to have a pre-deterministic effect (for new insights about turning on and off genes see e.g. Ledford, 2016), as it is known with cancer genes, they in general increase the cancer chance in a multitude of ways, and hence in general demand a greater regiment of prevention efforts to counteract them, both *ex ante* and *ex post*, and both at different levels, the biological and the individual level (cf e.g. Capellini et al., 2021; Roos et al., 2016; Byrne et al., 2023; Yu et al., 2023; as well as NCI, 2023).

> *Ideas, concepts, theories, cultures, accepted truths, accepted common sense, accepted methods of action, emotional and intellectual reflexes, social and cultural reflexes, etc., all act principally in the same way*, they can activate or deactivate (turn on and turn off) others, if they are programmed to do so.

The theory of human entanglements, thus, can and should learn a lot from e.g. cancer gene predispositions (probabilities) and their interactions with the biological environment (the human body and the human mind) and the language, social, cultural, physical and natural environment (all things said and done, all conditions, that are located outside the body), including all of entanglements of the past (past human conditions, past environmental, including cultural and lingual, conditions).

Meta entanglements are entanglements that control—activate and prioritize—other entanglements, such as religion or love (Nietzsche, 2014: 346–347, 2009a: 71) or political anchoring of one's outlook and identity (cf Holcombe, 2018, 2020, 2021), and hence ideology and attached ideas, rationalities and preferences.

9 Catalysts and Prohibitors

A long time ago, and it could have been yesterday, the finding is timeless, Émile Durkheim (1897) in his in-depth empirical study pointed out the importance of *social integration*, for the development or non-development of all sorts of social problems, taking the example of *suicide*.

We can conclude, hence, that social exclusion acts as a catalyst for suicide and other social problems; and social integration as a prohibitor for suicide/social problems, depending on the degree of their relative presence or absence. This is intriguing and still, as of today, offers a great deal of investigative venues and tasks to follow.

The search for other catalysts and prohibitors hence has just started. With the help of AI, new dimensions of massive data applications and analyses, and quantum computing, these forthcoming times indeed promise to be exciting (turbulent, but exciting) times.

10 Keys

In the theory of Niklas Luhmann (1984, 2012a, 2012b), and this is in fact very peculiar, it is referred to system codes, or key codes, they may be key licenses, documents, QR codes, passwords, literal keys, etc. When they are missing, they will cause a series of negative effects (or the loss of key statuses), e.g. job or job position. That is, then they also cause a series of follow-on negative events, ill health, divorce, loss of one's home, loss of one's children, and so forth.

These keys, as we may call them, are the guarantees and enabling forces of a great many capacities, which can also be understood as *powerful* entanglements.

They hold things in place, when, however, these *key vector forces*, as we may also understand them, disappear many other entanglements cannot be held in place and—as with a house of cards—things start fall apart, one by one, and rather quickly, in a causal chain of events.

11 Amplifiers and Dampeners

The environment, too, in the theory world of Niklas Luhmann (1984, 2012a, 2012b), which forms a great deal the basis for the Human Entanglement Theory, is causally enabling and causally disenabling in many ways, which are yet to be fully unearthed/discovered (cf e.g. Remington, 2023; Edin et al., 2023; Maine, 2017).

The environment, the human environment, the community, the local infrastructure, work and leisure options, exercise and sports opportunities, including the physical and natural environment, has its monumental influence by ways of holding key powers of influence, including absolute and relative degrees of possibilities, including their access chance and the degrees of convenience to access.

> The closer something is by, the shorter the time and the lower the costs to access it, and the more it was accessed in the past, the higher are the chances for it to be remembered, and to be chosen again, to be accessed again.

The geography of possibilities and opportunities is instrumental in also the way it not only influences us, but all people around us. It acts as amplifiers or dampeners of choices.

Sometimes this is due to role modelling effects. Often people choose to identify with role models, such as influencers, movie stars, the rich and famous, teachers, as well as e.g. deemed-successful neighbors, relatives and friends. They feel good about following them, while knowing that others also like or follow them; this is a double reward here.

Geography, however, is also influencing personal choices and behavior simply by communicating and showing the possibilities and opportunities at hand, among people within a lifeworld neighborhood or lifeworld region (valley, strip of land, etc.). In general, all lifeworlds (workplaces, settlements, villages, towns, regions) *also* facilitate the spread of communications revealing and certifying aversion/antipathy/hostility for someone who did not obey open or latent rules of communal/societal/cultural/religious behaviors.

Sometimes it is the (actual and/or potential) psychological yes and no of others that influences our behavior, in order to not stand out, not to be seen deviant in this or that situation, for this or that behavior, in this or that role. People are group animals, they like to seek safety in numbers in their surrounding peers, neighbors, family, relatives and friends (cf e.g. Habermas, 1981b; Clark, 2008; Remington, 2023; as well as Le Bon, 1894, 1895).

12 Turbulences and Warps

At times of upheaval, and due to yet unknown forces, human behavior is going through turbulences and warping effects of different kinds. *All—live, the universe—is not linear*, i.e. all that relates to and is related to myriads of human entanglements is not linear. This does not exclude the possibility and say anything, per se, about the likelihood of linear cause and effects, and their widespread occurrence.

Same with Kuhn's theory (1970a, 1970b), while there are sudden paradigm shifts, this does not mean absolutely that *all* is stable in the times in between one and the other paradigm shift, there are smaller shifts and movements. This is, to note, my take on things, and my conclusion due to my observation and knowledge of ongoing but rather slow scientific progress, also in a seemingly *relative* linear manner also in times of so-called 'normal' science (cf also Marx & Bornmann, 2010).

Physics leads with giving us examples, for turbulence and warping of human entanglements. This reference is to be used not as trying to find natural laws, or anything like deterministic predictions, from physics into social and life sciences. No, but as an inspiration, a creative source for contemplation and, in time, for theory building.

Divorce times are a good example of turbulences. Diseases, accidents, and falling in love are also good examples for both turbulences and warped human behavior, which can be the outcome of changed entanglements, or the cause that change entanglements. Sudden experience of riches, a lottery win, as well as the more often observable loss of fortunes, jobs, imprisonment, etc. can also serve as examples here.

I throw this up in the air (these balls), to indicate a way for further investigations in different fields, under different lights, with different means and objectives.

On the contrary to Luhmann (1984, 2012a, 2012b), I would like to give examples for each theoretical aspect of the theory under construction. He knew, wisely, there are way too many examples to give, way too many venues to explain. Yet, I have always wanted to prevent abstract notions without the reference to concrete examples, the more examples and applications the better. This is also the way I have taught Luhmann's theory of social systems, to all students taking a sociology class of mine, starting from first year students as well.

As a conclusion, here I need (and choose) to pass on these 'balls'—imperatives to investigate and to explain further—to others, to investigate each venue: be it a social, psychological, economic, financial (look at the stock markets, or currency markets, for example), political, biological, medical, geographical, or environmental, etc. problem or case.

13 Noise

When teenagers are encapsulating themselves, and trying to get distance from their parents and authorities of all sorts, in their *Sturm und Drang* period, they are untangling themselves from people for the purpose of being able to live outside in the outside world (to form new entanglements). It is tough out there, outside the family and the walls of comfort, the services and pleasantries of being at home, and not having to worry too much about things, life and how to sustain it in everyday contexts. They are programmed to learn to find their own ways, take their own risks, find their own creativity and solutions. New music, new idols, new look, and a new outlook and walk through life are part of it.

Psychologists described this as background 'noise', maybe it can also be understood (alternatively or complementarily) as disentanglements gone 'rogue' (homeless, shapeless, guidance-free), having lost their grip/anchor, being entangled themselves, hanging loose or running around wild, perhaps being entangled with or among themselves.

For sure regiment and outside (outside the person/inside the family) order, give security. But it also dampens the drive to creativity and adventure, the very things that are needed to take on the world, out there.

The lack of regiment has the effect of too many entanglements being loosened and detached, while others (other entanglements) have not had a chance, time, to form and harden out. It is a tremendous undertaking, growing up.

Yet, it is the very thing that keeps spinning the world, history, inventions, all that. It is the stuff that our life-stories are made of, as early on the most important decisions are being made.

I for one would not likely been able to write this book, without my most formative years, the late teenage years, where books and real-life adventures were the world that nurtured me, and they still do this very second, and the rest of my life, and beyond.

14 Pacifiers and Silencers

The history (our story) of humans is marked and pathed with pleasures all designed and functioned to hold us in place, to satisfy our senses and biological needs. This is ok, why not, this is good so. In any anthropocentric world, it is good for people, humans, to be happy and get what they need, what makes life for them decent, agreeable, comfy and cozy (*gemütlich* or *hyggelig*).

From *Göpekli Tepe* (Collins, 2014; Seewald, 2020), to Greek theater arenas, to Roman chariot-racing stadiums and gladiatorial arenas, to modern football, baseball, cricket, and all that stuff, meat and beer, or bread and circuses/games as it is coined after the Romans, or now football and hot dogs plus beer, beach volleyball and white wine spritzer, and so forth, this has sustained culture, politics, economics, it enabled and sustained the formation of civilizations (and contributed to their falls). It served/serves to pacify millions and now billions. It silenced/silences them, by making them content or even, for the most part, outright happy. It is a drug all have signed up for. A means to put an end to the thirst for other demands, of which (always) there are so many.

Modern, more stealthily hidden, versions of this include social security systems that people are supposed to 'enjoy'. This word is used so often—and one cannot overlook it, its design (purpose) and function, of the world 'enjoy'—*to pacify* the working and middle classes and *to shame* the ones that have nothing, or less than that, piles of liabilities, backlog of debts. Taxes are bought with welfare benefits, while the latter are also paid by the former.

Political pacifiers are all over the place, to bribe voters to get their vote, and support. Are (were) these not working, the punishing regimes (would) come in: incarceration of the masses for a mass of reasons, each coming with a pile of justifications/rationalizations; taxation, mandatory contributions, fines and obligations; court rulings (the rulings of long- 'explaining' judges); seizures of houses, cars and furniture; the kidnapping of children by benevolent-looking and benevolent-arguing government agencies; and this is just the start of the list.

15 Disentanglement

How to disentangle from intense alcohol consumption, or cigarette consumption? In general, any addiction, any entanglement, with a drug or another drug-alike substance (like sugar, cakes, ice cream, white bread/noodles/rice, beef steaks, sausages and so forth), is difficult to undo; untrain or reprogram, one might say.

Changing the environment, changing the environment of different times of the day, and of the week, helps a lot and/or becomes a necessary choice of action, as this is often indispensable to sustain a new habit, a new entanglement, hopefully a healthy choice of entanglement.

What is more, often, I would say—and NLP-fans would, too—reprogramming is key (cf e.g. Tosey & Mathison, 2010), by vigorously and constantly working on changing one's attitude and reasoning. Advertisements work due to their repetitions, for the very most part at least. The same is true with any kind of human programming, and any kind of thought formation, or idea formation. One needs to repeat the attitude formation and reasoning formation, just like walking over a wheatfield, one has to do it many times over to leave a sufficient/sufficing/lasting permanent imprint. The more repetitions, the more spotlight (sunshine) is thrown at the new mental formation, the new entanglement.

Another key element is the new entanglement, *a* new entanglement that takes the place of the old one (and time, in that same context), to make sure the old entanglement cannot come back (or has much lower chances, room, to do so). If possible, a new entanglement is helpful (and in most cases really needed), as the best way to fight an entanglement is another entanglement. A new habit, a new hobby, a new life choice, a new social environment, a new outlook in life, sometimes a newfound interest in spiritualty, religion or philosophy would go a long way to ward off old habits, old trodden paths of behavior and choices.

How to disentangle from a love? A new love. The medicine for love is love. I think, these one or two examples are enough for thoughts to spin further in these directions, independently (from this text), to solve the conundrum of not being able to give millions (or an infinite number really) of examples and to describe and explain them to a proper degree.

Everyone has a different take on things, and yet many things (entanglements) are identical across hundreds of millions of people. These is due to the (relative) *commonality of entanglements*, the common forces that brought them about, sustained and strengthened them all along.

16 From Social Systems to Human Entanglements: The Story

Niklas Luhmann (1984, 2012a, 2012b) has given all his (professional) life for the construction and refinement of one theory, his unique variety of a social system theory (that is not compatible, and not comparable, to other social system theories out there). The theory of his is also a theory of social communications, it can and is understood as such. I cannot compete with Luhmann on the matter of willpower, sacrifice and long-term commitment and planning. On the contrary to Luhmann and his theory, the establishment—or better final crystallization—of the Theory of Human Entanglement per se happened over a matter of weeks. It was never planned, or to be expected. It just came about.

Would I have not read a lot of Nietzsche not a long time ago, because I read Foucault extensively a couple of years back (whose source of ideas and methods was Nietzsche, cf Nietzsche, 1878, as well as 2014: 346–347), I would not have felt the possibility (or need, or wish) of writing a philosophy book, which now turns out to be a philosophy-cum-social-science book. A very unusual thing for me to do, a first of its kind.

Would Luhmann not have invested his lifetime in the making of his grand theory, too, this book here would have certainly never come to see the light of the day. But there is more: we need to go back further in time here.

The summer of 1993, in Tich Nhat Hanh's Plum Village in Southern France, was also vital. It was there where I learned about psychology from Tich Nhat Hanh himself, about Buddhist psychology (an independent and unique form of psychology).

Also, having read only, on and off, more glancing at them first, the two books my grandmother that she gave me for my 14th or so birthday, one was the Bhagavad Gita and the other were stories from the Vedic Scripts, two big fat books, this also contributed to the writing of this book.

The book vendor back then took advantage a bit of my grandmother, I guess, to get rid of these two books that would have been lost in that store for long time anyway, there in that remote mountain town.

There was no beginning, no one beginning. I started to look at the books first, the pictures and started to read passages here and there. These two books were not the reason why I traveled around the world for a whole year, not too long thereafter, I would have done it anyway, but they let me travel the world back in time, to a new culture and civilization. So, it did deeply affect me, in a very positive way.

My journey through life then also took me to a Krishna Temple in Penang, Malaysia, where I mainly was focusing on learning Taiji from a Chinese master (a student of the student of Master Cheng Man Ching); among other places. All of that, that chain of events, brought me here, writing this book, developing this theory. My thoughts connect me to those events through time, and place; as those two books connected me to ancient India.

Retrospectively, the *noodle (train) of events*, the *noodle of communication*, is easily detectable, visible, in this story. Writing the one or other passages these days on this book here, and reading those passages of the two books from my grandmother early on, these events, these thoughts and experiences are all connected.

So, back to Luhmann again, the formation of these two theories—*Luhmann's system theory* and our *human entanglement theory* here—cannot be more unalike as they are, one has been planned over a lifetime (Luhmann's theory), and the other (the entanglement theory) picked up along the way and invented randomly it seems, by chance—or better due to what *relative* 'fate' (the noodle) had in plan for me.

Luhmann invented a theory of *social* communications, and (along the way) I extended it, in the past, to also include intra-personal communications. This theory then had become (a concept of) a theory of general communication that I used for my own understanding of things, also for theory-making in my job as a social policy scientist. Now, with this book, this general theory outlook or general post-Luhmannian theory of mine, turned into a new—and of course fully universally applicable—theory: the *Theory of Human Entanglement*, which was created by way of entanglements itself (as everything is).

When reflecting the geneses of thoughts and experiences that went into this book, then, yes, this book, too, took more than 30 years to materialize, which is more in line with Luhmann's main work, his 1984 book on social systems.

The genealogy of thoughts, the genealogy of ideas, here, is not only useful in explaining the process and the origins of thoughts, theories, ideas, philosophies and conclusions, but also to learn more about entanglements themselves, how we are entangled, and how it plays out in real-life cases, with real-life examples (demonstrations).

Clark's strong insights and vision (2008) corroborate the *Theory of Human Entanglement* here and our theory corroborates his universal theory of the mind, the body, and the outside world. While Clark used different words, he referred to the same truth—that in a myriad of ways, constantly and ceaselessly, our mind, our body, and the whole world (social, physical and natural) are 'one unity' that is 'tangled' with each other, backwards and forwards, outwards, sidewards and inwards (cf also Phenomenon and thing-in-itself & Fundamental questions of metaphysics, Nietzsche, 2014: 20–24). The unity of the mind, the body and whole outside world is what here in this book is called the *noodle of communication*, which unites all three of these realms using their least common denominator, i.e. words.

17 Entanglements: No beginning/No End and What It Means for Social Research and Social Theory Making

Wars, addiction, abuse, neglect, inequality, a welfare state system, gender struggles and gender wars, failed states, culture failures, economic downturn and locked-in non-development, crime, violence and administrative/political corruption, all have this in common: they do not have one exact beginning, and one complete/exact end.

People's actions are not being called on or off, in or out. Life is not determined by doctors who call the time of birth and the time of death. The biological life is already there, and has already ceased to be before these were called. People's actions, collective actions and their consequences are not exactly (to the point) to be pinned down, with regard to their moment of entering existence and their moment of leaving existence behind.

Everything has a genesis—i.e. a history (a story) that led to the existence of anything. Everything has follow-on effects, consequences, memories, data, energy and ideas that live on, beyond what we call the end of this or that.

The same applies to entanglements, of course. These are formed, as anything else human and human-communication-based. A series of events, with an impossible-like likelihood, leads to anything and all things. The big bang, and what was before (e.g. a void, anything really), led to all human action. The aggregated likelihoods of all of the past that led to each and every human communication, and hence action, is not graspable. They are infinitely low.

The past likelihoods for anything that has come (materialized/occurred) have already carried, back then, in them, the chances and the seeds of the making of things to come. When your parents first met each other, that is already the start of the process of the formation of you and your life course, actions and thoughts (however impossible, or sheer impossible, they may seem at that time). And this goes all the way back to before, way before, the beginnings of what we call the modern human.

To give some examples, Switzerland and Sweden being/having been neutral countries rests on (and is founded in) their violent histories of fighting brutal wars, everything goes back centuries and millennia, back to early migrations, conflicts and human interactions of all kinds at that time.

Also, the first recorded labor strikes occurred under Ramses III, in ancient Egypt (GWR, 2023; Mark, 2017) on November 14, 1152 BC. No pyramid then could have been built without extensive systems of administration, employment and, yes, early forms of welfare state systems—these were already complete, in their way, covering the whole family, all major aspects of life, food, housing, medicine, education and more. These welfare state systems, temporary or not, were not the first ones. Wherever people lived together in bigger towns and large gatherings, life needed to be arranged, often (not always) this was accompanied by systems of common food production and food sharing/distribution, common protection against enemies, and different forms of mutual help (building houses/huts/shelter, caring and watching out for one another, etc.). The Vinca in and around modern-day Serbia, at their time, were prime examples for this, as were many other societies, from the Etruscan society and culture in Italy and beyond, to the Harappan society and culture in the Indus Valley, as well as many civilizations and societies in Africa and the Americas, in all of Asia and Australasia before that.

The main difference from Luhmann's system theory (1984, 2012a, 2012b) to this post-Luhmannian theory of human entanglements is that *entanglements do not have tangible beginnings and endings, the borders are 'fluid' and 'numerous and/or infinite'*.

This is nothing short of a *paradigm change* (from Luhmannian communication *systems*, to a post-Luhmannian world of *unity* of all sorts of communications that is created by, enabled and bound by, communicationally-constructed human entanglements). This is marked by the change from 'systems within systems' to 'everything is entangled with (in the end) everything'—in a unity flow of all communications (and communication-like actions), comprising all humans and their human-made communications (also in/of machines), flowing through time, thus, creating *the noodle of communication.*

Now things, as we may understand them here, flow endlessly intertwined with one another, everything is hooked in and based on (entangled with) everything else; to a lesser or greater degree, in this or that form, with this or that outcome.

There truly is no coming, no going, i.e. no (one) beginning and no (one) end; also for human entanglements. Everything, every understood truth, every perceived reality, and the universe perhaps as well, are all part of the same, *that what is (and it is and is not)*. They are united in ways we do not understand yet. And, for everything it is true that

> it is unbeginning unending (Parmenides of Elea, in SEP, 2020).

18 Knowledge Moves in Non-linear Motion

The ways of science are complex, and for those who are in the system all seems normal, as they operate in the Kuhnian (1970a, 1970b) '*normal science*' world that is controlling and feeding (handing out salaries to) them. While it is true, that is, it can be corroborated that science mainly progresses through paradigm-changing breakthroughs, from e.g. Newton's gravitational theory to Einstein's relativity theory, it is also true that there is constant sheer—but of course not absolutely—constant, progress in each scientific field here and there. Depending on the view, high above or low-on each-case, one can come to the conclusion that there are, also, here and there breakthroughs, again and again.

Empirically and logically, it is not advisable to support a conclusion that would say there is no (none) progress in between each paradigm change, within scientific fields. Freedom for scientists to rid themselves of the entanglements that hold them back, or threaten them with consequences (job, promotion, peer pressure, denial of publication channels, etc.), is always there and rather randomly (from a high point of view, that looks over millions of cases) distributed. And this freedom can rely on sheer randomness in terms of getting creative and path-breaking inventions, discoveries and breakthroughs off the ground. But, yet, in Kuhnian sense, there are visible and strong steps in the progress of sciences, even steps down, and not just up. Economics is a case in point (Piketty, 2014, 2015, 2020; Smith, 2017; Smith & Wilson, 2019).

While entanglements are structurally arranged through the workings of large and dominant systems and cultures, they are also, in their totality (all in all), ensuring degrees of random-like freedoms. Randomness, too, like the concept of borders, in human entanglement theory is not absolute, but relative. Rather, more random is the way to understand randomness, not complete or perfect randomness.

There is room for real innovative and all-changing creativity and freedom all the time. It is just that they are not dominant everywhere and all the time, but rather seldom and scarce. The Adam Smiths, Carl Mengers, and Thomas Pikettys of the world are not travelling in groups. There are no such groups, but only rare, and precious, individuals like them (Adam Smith, Carl Menger or Thomas Piketty). Greater and smaller paradigm changes are a main characteristic of the development of each science field, not just economics or physics. Relative longer, and on a close-up look also shorter, periods of creative stagnation (immobility) are the outcome of entanglements: including human institutions, as well as the cultural, professional (individual economic and social) realities.

All these entanglements are driven and composed of communications, abiding the rules and laws that have been set by communications, i.e. complexities and inter-complexities, and again *their* complexities, which we can all super-complexities.

Human entanglements in science development, are both prohibiting and diminishing in nature and/or compelling and propelling. They act *en group*, and united over time and (language/cultural) space.

They are located on different levels of human communications and human action, in the personal mind, in language, in law, in government and social practices, in social and cultural norms, in administrative and managerial norms, economic realities and incentives (negative and positive ones).

19 Randomness in Human Entanglement Theory

When it comes to theory, randomness helps us understand that all human action is not random, but for the very most part it is hooked up on strings, invisible and intangible strings, these strings we have come to call entanglements. They are forces at a distance (not a 'spooky force at a distance' as Einstein referred to quantum entanglement, '*spukhafte Fernwirkung*' is the exact phrase Einstein used; cf Einstein in Born, 1971: 158; Fine, 1996; Ferrie, 2023). They are forces of history, culture and language, forces of economics, politics, laws, administrations, forces of society, community and one's personal lifeworld (cf Gramsci, 2011a–c). They either prevent or hold back, or induce and push forward human communications, and hence human action. Nothing is completely random, not when it comes to humans.

In physics now, there is a new stream of thought (which the author is enthusiastic about, and fully supports logically and out-of-the-stomach kind of way, i.e. intuitively). This new thought has advanced the idea that all forces in physics are due to entanglements of quantum particles, different entanglements due to the different nature and interactions of different particles. Even gravity, and the space-time continuum are made by and subject to forces that have not yet fully been able to be put together in thought (logically, theoretically) and put to the test in any form practically. There is already a name for this new thought (theory) and that is *quantum spacetime.*

That is why when we talk, in theory or in methodology, about randomness, we are only talking about fuzzy randomness, or seemingly random events, as true randomness (the lottery kind of randomness) does not create and perpetuate human communications, and action. Every child's play, every next move of the child's hands, has a purpose, a beginning, and a learning effect from the past. Nothing is purely random.

In physics, so it seems, randomness is the lack of knowledge (Zeilinger, 2005) of what the plan (of everything, of the universe) is. It is the lack of knowledge, not the lack of plan (the dictionary meaning of randomness is lack of a plan, purpose and goal). What we do not know (yet, but some time in the future we will, hopefully) we like to call random. What is sure is that a long and steep learning curve is ahead of us, in physics, in all sciences.

Entanglements create the universe, what a revelation! Time and space are relative, and once called spooky entanglements (the forces of quantum entanglement) are absolute and real. If they were not, the universe would not exist.

> And yet, 'everything that is' is both '*it is*' and '*it is not*' (cf Parmenides of Elea, SEP, 2020)
>
> Anton Zeilinger strikingly has come to the conclusion that "the distinction between reality and our knowledge of reality, between reality and information, cannot be made" (Zeilinger, 2005, cf also Zeilinger, 2010).

The universe is coherent, as are all entanglements, the question is coherent to what, for how long, and to what extent, and when the coherence is changing/switching to yet another form and kind of entanglement/coherence. Every force in the universe is created by entanglements (coherence), by e.g. changing the environment new entanglements/forces come into play, new coherence (a new entanglement) replaces old coherence (an old entanglement)—what in physics is referred to when talking about decoherence, i.e. decoherence of the *old* entanglement.

Do we know *all* forces, *all* entanglements in the human universe, in the physical universe? No. Will we ever know?

Gödel (1931) knew the answer: *no one and nothing* will ever be able to calculate it! There will never be such an arithmetic, to calculate it all, any arithmetic will always be incomplete! It is and always will be impossible.

What we know is that we do not know *all* the forces, *all* the entanglements of either—human or physical—universe. We never will. And we do not have to know *all* of them. We now know more about them, we start realizing their existence, and we can now at least in the human universe begin to comprehend and hence steer and manage the forces (language, culture, laws, regulations, financial incentives, social and physical environment, messages, knowledge, choices, etc.) that create, prohibit and determine our choices of words, thoughts, feelings and actions. True randomness would mean chaos and no means to work for a better future. Fuzzy randomness can be explained, and we can work with it.

Human action, and human communications are not (truly/entirely) random, but they build on language, on common sense, on culture, practices and habits that have been established, that are amplified or curtailed, by human entanglements, like power relations of the Foucauldian kind, or any kind of monetary or legal relations, or any kind of emotional and logical relations. Human action and human communications may look random, and they are understood as random, as the forces that pull the strings in favor or against human communications (the very next word spoken or written, the next thought and feeling to arise) are not all known. Not knowing something in its infinite detail, and/or in its infinite dimensions and diversity, is not proof that something does not exist.

Chapter 12
Back to Quantum Methodology (Again)

1 Randomness in Quantum Methodology

As we have just established, when it comes to methodology, randomness is *fuzzy randomness* and not true randomness (the lottery kind). Fuzzy randomness is more or less randomness, it serves a *purpose*, to enable the answering of the research question. Randomness is not used because of a philosophical choice, and not for beauty either. Its beauty lays in its usefulness, the practicability of it, to contribute to and to enable a certain research goal. There is always a plan, the research goal. The choice of what comes next within the process of research, to be clear, is part of the process of one's research *work*, that shall be *reasonably* random, as it serves the purpose and goal of spreading out and capturing not just the obvious, the normal, the average case, example or phenomena under scrutiny, but also as many outlayers as possible—in order to find new gold, new eye-opening revelations and creative solutions, new ideas and new knowledge (cf Aristotle cited in Baumann, 2008).

Fuzzy randomness introduces a higher probability of creativity, which is certainly not a certainty. Due to higher diversity of observed cases, including higher diversity of levels and angles of observations, there is more clarity, which means a greater degree of wholesomeness (again we mean fuzzy wholesomeness). As in quantum mechanics, nothing is (exactly) absolute, all is to be measured in probability (cf Capellmann, 2021; Born & Jordan, 1925, 1930, as well as Oppenheim, 2023; Oppenheim et al., 2023).

Greater wholesomeness, not (perfectly) complete wholesomeness, will suffice when one looks for greater creativity, more capable solutions, new ways of understanding, a higher chance to enter and work on '*new lands*' (arenas/grounds) *of knowledge* (i.e. *knowledge that breaks new ground*).

C. Aspalter, *Human Entanglement Theory*,
https://doi.org/10.1007/978-981-97-5161-7_12

2 The Investigator's Need to Be Self-aware and Self-reflective

The general problem of using language and from being born into and thousand times shaped in different ways by our culture, education and social and communicational environments, can be counteracted with awareness and self-reflection by the researcher and scientist. Why they do the things they do, their motivations, their definitions, their bases of thought and reasoning, and so forth, all this can and should be scrutinized again and again, by personal and professional reflection (i.e. by way of reflection and contemplation of personal/individual entanglements).

Philosophy and theory of science is helping here, as they are made exactly for such a purpose. Thomas Kuhn's (1970a, 1970b) normal science phenomenon needs to be reflected in its entirety—the systems of publications, the systems at the university level, and academic level in general, that influence what and the way research fields, programs and questions are chosen, research methods are chosen, and how interpretations of findings are being reasoned, plus of course the channels of publication sought (from the very onset of the publication process, advisably).

3 Empirical Theory Needs Empirical Input

There are different kinds of theories. From the point of testing a theory, there are falsifiable and fallible theories in the social sciences and humanities. There are purely reason-based, abstract theories (e.g. theories of parallel universes) *or* empirical theories, like Luhmann's, or Freud, Adler and Vygotsky's theories.

All are valuable, if they are valuable, that is, if they can inspire and propel knowledge creation, deeper and wider understanding. There is always a test that either corroborates, substantiates or negates, falsifies a theory. The test may be in a falsifiable nature or in a competitive nature (with other theories, and see who is better, and only the better one wins, all others are to be sorted out).

Vygotsky's theory for example is an empirical theory, we can easily observe young people, adolescents, children, infants and babies in their environment. We can measure, double check and corroborate. The same is true, with the human entanglement theory, one can test it in a thousand different ways.

Take finance for example, Joseph Alois Schumpeter emphasized the importance of credit in business and economic development. One can observe, measure and analyze credit applications, approvals, and their amounts and interest rates, and other conditions, and one can measure what this money has been used for, and measure again the outcomes (market shares won, new technologies applied, new products developed, new companies acquisitioned, etc.), and compare all these with corresponding measurements and outcomes of competing companies.

We are analyzing, then, networks of financial entanglements, market position entanglements, technological entanglements, productive entanglements and power entanglements.

In biology, too, one can study human communications, and take advantage of human entanglement theory, as we can look at neuron networks as they are producing words, and words keep running the human action world. The biological brain matter, the biological world, brings forth the communicational world. Neuron networks are also representations of entanglements, even on a much greater scale and complexity that can be truly fathomed by the non-expert. Both biology and communication theory can learn from each other. The ways we produce sentences, connecting thoughts, within the neural networks of the brain, may shed a great deal of extra light onto the making of human entanglements and the storage and interactions of the very same, on micro-micro level (on the atoms of human entanglements which are compounds of neuron activities and processes, e.g. thoughts and feelings produced, stored and interpreted; memories stored that are queried and constantly reinterpreted; words that are associated with different conscious and unconscious meanings, and so forth).

This new century will prove to be truly exciting with what science, all sorts of sciences are about to bring forward. The empirical needs to support the theoretical and the theoretical needs to enable the empirical. Theories are made from the empirical, out of the empirical and are based on the foundations of the empirical. And theories are to serve the empirical, to be useful in the real world, for science and for humanity.

4 Follow the Lead

It is simple, the technique in general of a crime scene investigation. One has to follow always the lead, the empirical evidence. One however needs to probe constantly the correctness and appropriateness of the working theory or theories, and to decide to dismiss or re-rank their likelihood and/or usefulness.

A perfect example for crime scene investigation is when one is or wants to mine a salt mine, gold or opal mine. One has to have a look at the experience of other miners and mines, and look at the landscape (of people in the case of social sciences). There will be guiding theories already set up by others, they will contradict one another, and/or support one another. One has to carefully study them, and decide their value and likelihood of correctness. One has to apply both of Steiner's (1988) razors, the first razor *is it a theory at all* (despite random name-calling, as it is also and sill rampantly common nowadays), and the second razor *the best theory needs to be supported while all others need to be cut*. For Steiner, only the best theory—no matter by how big or how essential a margin(!)—is allowed to continue join the competition of theories. All others need to be discarded, by the investigator in question.

Success needs to be sustained, by following the lead and updating constantly all the evidence and theory evaluation.

5 Attention and Experience

Attention is caring, experience is power. Both are vital ingredients for the investigative method in science. A thorough investigation demands full attention to all empirical facts, all data available, the ones in the spotlight and the ones not yet sought and found, as well as attention to all knowledge found and all knowledge that is yet hidden. With longer periods of experience, with longer periods of quality (eye-opening, brain-exercising) experience, the process of investigation becomes in general more successful (when taking a probability perspective). The expert dancer does not learn his or her skill overnight, time and training are of essence, and so it is with the scientific investigators/researchers in each specific field and/or across *different* fields, which is arguably (and experience-wise) more difficult yet again—i.e. transdisciplinary research investigations.

6 Picking Up the Lead with an Open Eye and Open Mind

Tread lightly and do not follow others, *just because* they are or think they are important (or at home at important institutions, universities), or *just because* they form a majority, even a clear and overwhelming majority. The lead is important. It should be one's own lead, or one of another that is deemed worthy, i.e. being the best in terms of one's own judgment and experience (based on their past success, and promising future successes). One has to abandon one's lead immediately if necessary. To err is not a shame, to not admit it *is*.

Every pioneer is to make difficult choices, for were it not difficult, one would not be a pioneer, by definition and nature of being a pioneer in general. Pragmatism in e.g. a practical science, like social policy science, is key (cf Pierce, 1878a, 1878b, 1893; William, 1970, 1981; Dewey, 1910, 1922, 1927).

One needs to be ready to switch, adopt new ways of thinking and working, to reject and to support old and new ideas, based on evidence, past and current successes, and good old intuition (which is in fact always an amalgam of one's past experiences and one's current analytical assessment and thinking).

7 The Shepherd of Knowledge

A scientist is not a worker bee without a hive, not a shepherd without sheep or other animals to guard. A scientist always needs to keep knowledge creation in mind, to enable it, to protect it (from the forces of prohibitive or distortive entanglements) wherever, however, and how much as possible.

Holding back on new knowledge creation (Conant, 1957; Kuhn, 1970a, 1970b, 1977, 1978) is hurting knowledge, science, the very things that needs to be protected.

James B. Conant (1957) was the first to emphasize the behavior of scientist to support old theories and methods, and to disregard new theories and new methods for the reason that they are new.

Discomfort (when one has to step out of one's established comfort zone), *procrastination* and especially *protectionism* of one's own theories and one's own old knowledge, or that of one's idol scientist, and (by now) outdated state of the art theories and methods are the combined reasons for this, and they are to blame for the sluggish, held-back and locked-in states of scientific developments that were and are all around us.

8 Comparison as a Base for Theory Development

In the social sciences, it is vital to use the method of comparison, as the fundamental instrument to come to knowledge, and to test one's knowledge, again and again, all over the world, in each subfield or perspective of application.

The widespread relative, and often also absolute, and even in its entirety, absence of theories (*'real' theories*, and not the ones that call themselves theories, but are not, cf *Steiner's first razor*, Steiner, 1988) is more often than not owed to the lack of comparative research, i.e. the lack of empirical analyses, including case studies from all over the world, in every corner of the world, as well as including data about all people, their conditions, their institutions and countries in each and every (if possible) country, and its regions, of the world. To give an example within the wide range of social sciences, let us look at the majority of normal science (Kuhn, 1970a,b) that still exists in the social policy science:

> The ("*in a way* medieval") old ways of looking at one's own neighboring countries and that is all, are outdated—that is, made completely obsolete by the wheels of time, of world history. A rising number of professionals and academics of comparative social policy have already come to see and understand that there is only one world out there, one playing field for all of comparative social policy, and that includes all countries in Africa, all countries in Latin America, all countries in Asia and beyond! … We need to look at all countries and regions of the world, we need to look all fields and subfields of social policy—in comparative social policy—and bring them together, building *the puzzle of social policy*, that is prone and subject to ever more challenges and culprits of our time in the near and not-so-distant future, that are either highly visible and deeply felt, or completely hidden and/or ignored. (Aspalter, 2020b: 253–254)

Chapter 13
Keep Spinning the Wheel: Looking Again at the World Out There

1 In the Zone of Mindfulness

We do not have to become all *Buddhas*. We do not have to say *no* to all the things in life that we enjoy, not at all.

Enjoyment comes with entanglements, getting used to it, wanting more, etc. This is ok, all we need to be is to be more often—the more the better, not perfectly, and not *all* the time—mindful of us, our body, our surrounding, our society, culture, our past and future.

2 The Source for the Art of Everything: Inspiration

The quantum method of looking for and logically and experientially substantiating the theory of human entanglement has led to a major conclusion; one that is precious and at the same time easily overlooked. That is, working with randomness, working in parallel, and working with the universality of seemingly very different things have opened the door to many new endeavors, many new findings, and the most persistent force behind all these is *inspiration*. The quantum-inspired ways of conducting social science and philosophy theory development have manifold increased the degrees of inspiration. It also prevented the forces of path-dependencies of other, much-trodden paths and methods, plus old and thus not-up-to-date methods and theories.

Science, theory-making, knowledge, and of course wisdom, are forms of and the outcome of art. Being inspired is an experiential art. Almost no scientific breakthrough ever came without it. Not in the case of Albert Einstein, not in the case of millions of other scientists. They all needed, the all need *inspiration*, *imagination*, and the *dreaming* (the spinning of the past into the future) that leads to future scientific breakthroughs (cf Schrödinger, 1935a: 807–808).

C. Aspalter, *Human Entanglement Theory*,
https://doi.org/10.1007/978-981-97-5161-7_13

3 The Purple Gold

We are looking for gold everywhere, for the wrong kinds of gold. Data mining is the new source of power and money creation (absorption from others). We are still just chasing old traditional gold.

We have not evolved into creatures that seek wisdom and knowledge first, not even health and happiness first. We are primitive, as can be. The well-being of our elders, the laughter of our children, the organic vegetables in our garden, the super foods on our table and plate. We must come to treasure foods in super-rich polyphenols, like purple sweet potatoes (purple *batatas*), like super-fiber-rich oatmeal. We need to aim for the world population to get apples, bananas and oranges (and more) every day in their lives, to have a good, and decent, life. To live healthily and not to make each other exhausted, anxious and tired to the point of missing out on all the wonderful and really important things and events in life.

4 Pride

It can be a force of encouragement, a force that provides the essential ground for confidence. Any parent knows to support the pride of their child, otherwise there will be no confidence and strong decision-making later on in life. Well done, great, we say. This is needed also with adults. This is the positive side of pride.

The same force, however, as we already were thinking while reading (me, writing) the last paragraph, is also very destructive and impeding to human progress and the development of human decency (well-being). Too much of something good, is called a poison (medicine turns into poison by way of increasing its portion). Policymakers have not yet focused on the psychology of human being, marketing experts have. Medical experts, social policy experts, social work experts, too, still have to come to see all the positive and all the negative forces and instances (contexts) of what is understood by, and entangled to, what we call pride.

5 Prejudice

Our human brain is better than any machines made so far, as our human brain works efficiently, cutting down on data processing, in a great many ways. This is like calculating with stereotypes. Big cats are dangerous. That was our experience, and this formed stereotypes and prejudices, and bigotry all the way, with regard to thousands of other things. We train ourselves to form non-realistic stereotypes, like elephants are nice, they are cute, but yet they are as deadly as big cats (lions, etc.).

Prejudice, is as the word is already saying a pre-formed judgement. A shortcut. Hence, we are working efficient. But this has many downsides as well, we need to know our past, in the Savannah (that one). Where we come from, there, this was an urgent, most important, need to stay alive, and this we had achieved in the past by using prejudice and stereotypes. This happened as we still were pre-human apes, when we have turned into *homo erectus* who conquered most parts of the earth, all the way through the Ice Age and up and beyond the Middle Ages.

We are still like this. So, what to do? Reflect and move on, try to accommodate to our nature, understand our human past, and our current conundrums of how usually, for the very most part, our human brain works, which leads to imperfect and wrong human communication and imperfect and wrong human actions. Only in this way, after we come to understand and accept this, we can start to form strategies, policies and systems to change our culture of everything, bit by bit, in a human-respective and all-caring manner, ethically and compassionately.

Otherwise, if we do not do this, we will fall back into the loops of war, revenge, mass abuse of power, mass dehumanization of all others, in the many—some overt, but mostly stealthily hidden—ways of human entanglements.

6 Running with no Goal

Mental formations, stubborn ideas, and cemented stereotypes; publicly-, politically-, socially-, culturally- and linguistically-institutionalized ideas; cultural and social power relations, as well as logged-in economic development paths (plus situations, and their conditions), they all are created and case-hardened by human entanglements.

These work in ways, and are created so and managed so, that they keep us in place. The powerful ones in ever-more powerful place. The powerless and disadvantaged in ever-more powerless and disadvantaged place.

All of us, so it seems, are running on the wheel of humanity, created by it, its existence, and its inherent forces. These forces propel, compel, twist, obstruct and destroy. In total, this infinite number of forces (entanglements) is having a real, finite impact on us in terms of their aggregate power; but infinite in terms of timeline.

These forces are *not* all-compelling, all-triggering, there is room for plenty of freedom and self-determination—as any human world must have, for humans to remain humans.

Yet, due to contentment, getting used gradually, getting tired over the years, etc., the vast majority of us come to *accommodate* in the long run a great many, if not almost all, *disadvantages, limitations, exploitations and traps*; as well as their made-to-be-convincing rationalizations, plus their fermentation and cementation in our minds and culture. We are by and large but running all of our lives on a big giant wheel, that is made of and contains the universe of human entanglements. Our next generations will continue this 'work' and 'existence'.

7 Nietzsche's Ladder of Life

Life is sweet, life is bitter. With every year, we grow older, and more experienced—in terms of both, good and bad experiences that we have made and encountered.

The older we get the worse is our health, if we do not hold against it (fight it) with all the forces and means that we can get to know and can get hold of. The older we get the less we have opportunities in life to change our life onto a different trajectory.

On the other hand, it is also true, that the older we get the more we can achieve in many different aspects, be it economically, socially or spiritually. And, the older we get the more we have the chance to nurture and grow our wisdom and knowledge. We are descending the ladder of life, as we are ascending it at the same time—both at different speeds, and in a non-linear fashion (cf Nietzsche, 2009b: 74–75).

8 Injustice

It happens so often, and yet it is being talked about so little, and realized and paid attention to so little. The rich and powerful get a slap on their hand when the embezzle millions or commit tax fraud amounting to many millions. All others, i.e. the little people and normal people in general, go to jail for nothing or extremely little things, and for long times for small offenses and even non-offenses, just non-conform behaviors.

The systems of injustices are formed, carried and protected by the thought systems and Foucauldian power relations (Foucault, 1975, 1976; Lukes, 2005; Rabinow, 2010) that are apparent and exert their forces only in certain hierarchical and institutionally and culturally long established (and solidified) relationships in between different kinds of people: the rulers and the ruled. The rulers are the people in power, in position, with a license of power over others, the power to imprison, penalize, discipline, tax, judge, evaluate, shame and blame all others. And, the rulers make sure, for themselves, not to have to face and answer to these very same systems of control and oppression themselves.

The mortar and the bricks that build these 'power relations' are *past actions of communication*: systems of rationalization (that weaponize logic); arguments and reasoning of moral superiority and God-given or God-given-like institutional authority; arguments and reasoning of providing security, order, health, well-being, protection and support; arguments that justifying death, war and violence for the sake of peace, prosperity, a better world order, or a better *Weltanschauung* (worldview), for religions and cultures, for a faint prospect of realizing or reaching this or that political dream (a socialist utopia, or the one-out-of-a-thousand-reachable American dream).

The communications that have built the power relations that have grown over centuries and millennia, are institutionalized in language; ready-kept ad hoc explanations and answers; common logic and common habits of the people; common culture and preferences; political, administrative and legal institutions and practices; systems of social and economic control and suppression; rigid (and often totalitarian or totalitarian-like) religious practices and other traditions, and, last but not least, systems of media that fan the fire of fear and hate by the commoners, and their admiration and support for their dominating elites.

The constituting divide between the ruled and the ruling castes is formed by wealth, wealth of the past generations that is passed on to next ruling generations of the super-privileged, the super-rich and the super-powerful.

If we want to tackle injustice, the only way to do it, is to give back the people of this earth the wealth of this earth in form of taxing the super-rich, e.g. everything above a certain threshold, or the very most of it (Aspalter, 2023a).

9 Revenge

An eye for an eye, this is still an evil recipe that is applied all over the world, by gangsters, by gangster politicians who turn their nations into gangster nation states. Many times, it is interpreted and practiced as at least ten or twenty eyes for an eye. Things are getting so much worse; never better, one may conclude. It gets all rationalized, street cred and notions of gangster respect here, national interpretations of history, notions of peace through war, ideas of global world ranking/order there.

10 Sehnsucht ('Yearning') as an Addiction

The German word *Sehnsucht* literally means 'yearning addiction' and gets the point what yearning really is, an addiction. We want money, fame, power, and to be important. We are all driving by these forces, of yearning. The German word for addiction *Sucht* also includes the word 'seeking'. Seeking is an addiction, if it is done constantly and where it has reached a level (in terms of time and influence) that crowds out other important things in life: the love for family members, the love for oneself, the protection of one's livelihood, etc.

Words are sometimes so telling, and most of the time we are misguided by them, as their denotations and connotations are full of historical, cultural and common-sensical entanglements that are created by yet bigger forces of entanglement (inequality, hierarchy, poverty, gender roles, cultural and religious practices, etc.).

11 Winter Is Coming

For people in the subtropics, winter is a good thing, it is the time to harvest strawberries (instead of rice), to enjoy the cool weather, providing relief and sanctuary from the long hot and brutal summers.

In the north, where I was brought up, things are entirely different, survival depended on sufficient, soon enough, and successful preparations. Cutting, carrying and piling wood in summer was mandatory. And so was filling up the cellar with marmalade, compote, fruit juice, vinegar, cider, a dozen or so crates of apples, cured and smoked pork, pickled cucumbers, dried plums, Sauerkraut, i.e. literally anything that can be stored and eaten or drunk later.

Later summer and early autumn are used to go into the woods, to get elderberries and mushrooms. I remember, my grandmother once took us out to harvest chamomile tea (the right kind) from the paths in between the fields, to get more, we did have quite a lot of them too in our garden on the edge to the street. The garden provided food deep into autumn, with Endive salads and apples coming in when the first snow was falling already. Nothing was wasted.

Now, supermarkets and their prices are making everyone poor. Everyone is helpless, the woods and gardens are abandoned, for the very most, by the very most. Capitalism and its promises of individual careers after long years of (heinous) education for the purpose of independence and a better life, have resulted in *dependence* and a *less-well* life.

Economic slavery for the rich elites that own and run electricity companies and supermarkets has replaced economic slavery for the old aristocrats. New aristocrats with new faces and disguises have replaced old aristocrats. Pricing systems and the workings of inflation on the income and wealth of the common people have become their new blood hounds.

12 When Disaster Arrives

It happens again and again, every year, in some parts of the world. Disaster strikes, and nothing stays the same. Modes of feelings and thinking change, priorities are reshuffled in an instant. Outlandish scenes, horrible scenes are taking hold of one's attention. Scores of entanglements are pushed out of sight or tossed out, others are amplified and new ones installed.

13 Bigotry Breeds Hate, Hate Breeds Bigotry

When hate is used as a means to fight bigotry, it will most certainly defeat its purpose, to defeat bigotry. Quite the opposite will be achieved, that is, bigotry is strengthened by the hate it receives, and this breeds yet more hate, and the cycle of hate and bigotry is complete. Reverse-bigotry is bigotry. Responding to hate with more hate creates just more hate that breeds more bigotry, on all sides.

14 Selfishness and Jealousy

We were not taught in the right way, when we find that we are plagued and tortured by our own wrong behavior and decisions, especially when they arose out of selfishness and jealousy. The entanglements that push us to pursue selfish thoughts and selfish goals, to feel envy and resentment, are the ones that were passe on to—established, amplified and solidified—by our education, our practice and experiences. When looking at it from above, from the outside, our lifeworlds, societies and cultures gave them to us, to our parents, to our peers, teachers and neighbors, and they again gave them to us.

We also participated with our choices of what to follow and what not, what to follow again and again, enthusiastically or not. We suffer from this, as we have made and make the world around us suffer.

If we realize that selfishness is not in the best of our interest anymore, if it got out of hand (way out of proportion), if it became a destructive and hurting and hurtful force, it is time to let loose, to let go of selfish aims, and jealous perspectives. It is not too late, it never is. This applies to people, and the very same to communities and countries.

15 Compassion as Meaning in Life, and Source of Creativity

On the way forward, in society, we need to learn to learn. A necessary precondition of learning—in society and for society—is compassion, the ability to walk in other people's shoes, to feel other people's disadvantage and pain, their worries and devastation. Compassion is the passion one can generate for others, when one is with the other, in person (physically) or in mind (mentally).

We—if we want to progress, and stop all the regressing, all the regressing, or most of it, that is going on everywhere—need to teach our young to think compassionately, to learn to feel passion for themselves and for others, for all others.

Society needs compassionate citizens, and compassionate leaders to lead and take care; physically, emotionally and financially(!). Greed is decapitation of the soul. Compassion is the making of the soul.

We can learn, we can leave the specters of past and current idols behind, we can look for new creative ways to follow, new creative ways to be whole, i.e. complete, decent, happy and humane. Indeed, we can.

All entanglements can be dis-entangled, and re-entangled. In the words and principles of quantum physics: decoherence is followed (brought about by) by new coherence. New coherence causes decoherence.

The *environment* is key, and *communication* (the force in between/the force that entangles) is key. Once we get these two right, coherence will be reestablished in new, and lastingly positive ways—if we choose right, if we manage to make better decisions, and establish better ways of learning; living and loving.

16 Seizing True Happiness However Much Possible and Warding off Everything Else

It is true, it sounds too good to be true, but it is true. But it is also simple, not too much to it, not really. One does not have to negate happiness, one can enjoy happiness, seek happiness and be oneself truly entangled with it, and by it. As long as, one feels good about it, one shall stick with happiness. Perceived happiness and real happiness, however, are very different things. The one does not last, and makes oneself or others around oneself feel pain, sooner or later, and the other lasts and feels good, for as long as it lasts, and for everyone.

Like in accounting, the positives are to be accumulated and guarded, appreciated and sought to be made permanent and multiplied. The negatives, on the other side, need dissipation and expulsion, steadily and constantly, and energetically so all the way.

Chapter 14
The Non-final Curtain: Conclusions Derived from Human Entanglement Theory

1 Our Mental Corset, Our Tainted Glasses and Our Luggage on Our Shoulders

It can be tied strongly, and super tight, and usually it is—our mental corset. To loosen it, requires years of training, as ancient Greek philosophers like Socrates, Plato and Aristotle, new.

We carry with us not just luggage on our shoulders, our own past, that of our people, that what we have been told (indoctrinated really, every telling is indoctrinating) in our school books, in our newspapers, on our television, on our phones.

Also, we have glasses on that are heavily and, in many ways, one-sidedly tainted, misinterpreting and twisting things to fit our formed preferences (established mental formations), finding reasons to hate and distrust others who have, portray and defend different ideas, solutions, and ways of looking at and thinking about things.

As if this were not enough, we can say that the presence of us-changing entanglements that we carry around with us, our assortment and collection—that have been given to us by communication from 'the outside world' in Luhmann's terms (1984, 1997), 'the people we are connected with most' in Vygotsky's terms (1978)—also represents a mental corset, one that is fastened extremely tight (in quantum physics, there is a very similar principle of *loss of coherence, or decoherence*; the point decoherence occurs when the environment starts to exerts its own influence over the observed object/subject, cf Arndt et al., 2005).

With it we react deflective when others are trying to enter and/or play on our playing field, our occupied ground, our expertise, our field of practice or profession, party, country and culture. We react strange, suddenly estranged, we are compelled to be carried away by our over time hardened entanglements (over our lifetime hardened, starting from earliest childhood, cf Vygotsky, 1978).

When certain things happen, we are pushed to fall into despair, into anger, by forces beyond our control, that are so close to us, so deep in us, that we cannot see or comprehend.

C. Aspalter, *Human Entanglement Theory*,
https://doi.org/10.1007/978-981-97-5161-7_14

On the other hand, we are drawn to well-receive certain things, be enthusiastic about certain things, crave and cheer on this or that, these are also compelling forces. There are obstructing and compelling forces that we carry on us and *deep in us*.

> These entanglements are holding us up, have formed and directed our entire past, and they provide only room (freedom) for this or that maneuver in this or that direction, when this or that happens, for this or that envisioned and sanctioned attempted goal and purpose.

There are positive and negative entanglements, and what is positive and negative, i.e. understood and perceived as such, varies from different viewpoints that have been programmed in the past.

These things and their reactions in us are *pre-programmed*, by our past emotions, past thoughts, and past experiences, having received all the entanglements that the world outside had prepared for us, and delivered to us via our parents, friends, relatives, kindergarten, school system, legal system and court rulings, rules and regulations, punishments and rewards, in our workplace, in our community, as well as dialect/language and culture.

We are standing too close, we are too attached, too immersed, we cannot see the whole picture. Not the reality of Plato's Cave (Plato, 2011: 514a-521d), and not Weber's *stahlhartes Gehäuse*/Iron Cage (1952: 181–182), not the reality of mental Forms (Plato) and mental formations (Tich Nhat Hanh, 1993), we are not programmed to see them, we are supposed to work below that sort of radar and understanding, for the past tens of thousands of years at least.

> Our mental formations, Tich Nhat Hanh says (in his teaching that lives on), are like a garden in our subconsciousness, filled with positive and negative *mental formations* (cf e.g. Hanh, 1993). This is what I understand as *entanglements* in the Theory of Human Entanglement.

We have locked up this garden, in the back of our mind (brain). But, we can tend to it, by tending to our thoughts and emotions that have, of course, ways to penetrate the deepest of our subconscious parts of our brain. The biggest problems, negative mental formations can be softened and made to disappear/dissipate (slowly) over time. The more we focus on them, the more they grow; we need to embrace them with love and compassion, but not increase their water and sunshine—i.e. the attention they receive, e.g. by anger or despise towards them, or trying to avoid them. Giving attention to negative mental formations, entanglements, means letting them be in the center of our emotions, and constantly and randomly invade and hijack our mental living room in our brain (e.g. cause a scene, a firework of compelling emotions, or violent outburst, etc.), the conscious part of our brain.

2 Forgetting and Disregarding: The Open Secret to Useful Knowledge

The way things are is subject to a logic that emerged in and was tested over the past, our long, long past. That we forget things, some easily, some things rather easily, i.e. that we forget things in general, is a progress, an economizing addition on our track of human evolution. If we were to remember everything, we could not pay attention to the danger of the here and now, the beauty of the present, right here in front of us, besides us, all around us. We need to be able to forget to reach a higher purpose, a higher understanding, and to be able to stand on the ground, in the here and now.

We have all experienced, to some degree or another, travelling with too much luggage is not practical, it is hindering and counterproductive. Hence, it is good so the way it is, with our human trait of forgetting things, of pictures fading in our memories; luckily, we have cameras now to hold onto some so precious things that we want to revisit in more clarity again and again.

3 We Need to Be Nurturing the Human Soul by Nurturing and Grooming Communication

There is a normative conclusion to all of this, the theory of human entanglement. As an optimistic person, as in general I try to constantly install and keep up that momentum, and this was given to me by the teachings of Krishna, Buddha and especially Hanh (1993), I came to understand that everything can be fixed.

Everything can be repaired, as everything is made out of communications (words, thoughts, feelings, ideas). Whatever has been *can be amended*, just as Tich Nhat Hanh has proved to us, and what he has taught and is still teaching us (through his books and recorded lectures).

This conclusion is of monumental importance for policymakers, for psychologists, for development economists, for medical experts, social workers, philosophers, and for all of us.

4 Social and Economic Problems, Too, Are to Be Met with Strategies and Solutions that Focus on Communication

For long, I have been propagating this already, communication is a means to do social policy, health policy, and kind of public policy. Using marketing techniques and strategies, and communication channels, to enhance people's awareness and to enable thus their free choices is the way to go—and not nudging or pushing them, that is entirely unacceptable as it robs people freedom and self-determination, and hence of their decency and humanity.

This book here serves as an advanced (on the level of philosophy and all-encompassing inter-human and intra-human interactions) and extended introduction to the mechanisms of communications and the forces that are built with them, which tangled up our worlds of communications; having entangled our every thought and feelings, our ideas and aspirations, or evaluations and preferences.

5 Language and Rational Thinking: The Greatest Inventions and the Greatest Sources of Problems

Contrary to Habermas, it is proclaimed throughout this book, as before (Aspalter, 2023b), that rationalizations and declared rational thinking can *never* be free of the forces of history: the history of culture and the history, meaning and connotation of words, situations and contexts; as long as we use words we are tangled up in a web of compelling and/or obstructing forces, the web of all-penetrating, all-affecting human entanglements.

Language and rational thinking—as Nietzsche (1887) has already demonstrated so in a wonderful way before—are key sources of problems. Much earlier still, Confucius (1893) talked about the strongest conundrum of all times, which is the giving of names, which divert our thinking and our human actions. Wrong name-calling, wrong name-giving in the first place, is the source of thousands of problems in our daily lives, all our lives, since as early as humans managed to utter words, or at least not too long thereafter.

6 Everything in Communication, Everything Based on Communication Is Entangled in a Myriad of Ways

There is work to do. We have a lot of work in front of us, if we want to dissect all the entanglements, all the causal connections, that shape and force our human communication, and all human actions made out of and based on these communications, including the spun-on human communications that now exist in form of human-to-machine, machine-to-human and machine-to-machine communications.

Nietzsche, now, and probably, would send an invitation out to all scientists and experts in all the sciences and disciplines of language, history, communication, psychology, sociology, economics, political and policy studies, and of course philosophy, *to board the ships and venture out* into different directions, different oceans and seas, to all corners of this earth and all universes of knowledge and expertise one is familiar with (cf Nietzsche, 2009a: 163).

Everyone to their needs, wishes and capabilities, no need for rush, we waited thousands of years after Plato (2011: 514a–521d) told his parable about the cave, some more years, some more days, we will endure.

But, silently, there always is a competition going on among scientists, and full-fledged philosophers are among these scientists. And, some do break out of habit, out of the well-established, well-guarded comfort-zones, and some, here and there, break up and propel the current state of science, moving on in a non-linear fashion and in positive, towards-meaningful-results-pointing directions.

As Kuhn's theory (1970a, 1970b) predicts, there is always room and chance for a paradigm change, and a change of culture in science and theory formation, theories are knowledge, science is made out of knowledge, i.e. theories, and nothing else.

We need more knowledge to be set up and tested, empirically and logically, and in a recurrent manner.

7 We Still Have a Long Way to Go to Fully Understand Human Action and Communication

There will be soon revelations of now unimaginable sorts and consequences. That will not happen too long in the future, but rather in a couple of years from now, with the ascent of quantum computing and quantum communication research, and technologies applied, of all sorts.

It is hoped, that we can guide, within the limits of the bounds of this book, this process, or at least accompany it, as an observant, a willing-to-learn observant, as scientists, commentators, or thinkers of all sorts.

What is to come will change our lives. And we ought to prepare, what we already have set in motion. The technological capabilities beyond usual grasp are just around the corner. It is wise, and it will prove to be useful (and indispensable), to harness the power and potentials of these new technologies for all of humanity, and ward against the tremendous risks they bring, and their tremendous downsides there are, the problems that come with it, and are exacerbated by it.

We need to learn to speed up our philosophic understanding to keep track and to keep up with the latest and soon-to-be technological developments and *faits accomplis* (vollendete Tatsachen).

Having said that, it is also true that quantum technology is in the position and will function hence to serve as an extremely useful tool for research, and fostering the understanding of human decision-making, action and communications. There are long ways to go, and patience will be key.

But philosophy can be (figurative) light seconds ahead of all this. We can predict, we have the power to predict, those findings. We can imagine, dream about, think about and think out, all these entanglements and in parallel start the testing, also in parallel start our analyses, our interpretations and ever-better prescribed or offered solutions.

Both scientific enterprise and philosophic enterprise of the scientific enterprise, are processes, they will take their time, as usual. They are not one-time acts. There are acts that are all connected to each other in parallel, and back through time, into history.

8 Quantum Technology Will Soon Facilitate Predictions of Human Actions and the Consequences of Quantum Technology Itself

Looking into the future is amongst the most powerful things to do, only being paralleled e.g. by looking into human brains and the many details and revelations of the human DNA.

With the rate of increase of computing power, and especially predictive computation, that has been already set in motion with new breakthroughs in quantum computing technologies, we look at a behemoth of changes—in other words, an extremely large and powerful set of changes.

Not only need we to predict and measure, more or less, the magnitude and speed of things to change, but also their impacts, the magnitude and speed of their effects, side-effects and follow-on effects, all the unfathomable many of them (non-finite many of them).

For this we can, and hence we have no other means anyway, and why not, try to use quantum technologies as a tool of prediction and a tool of speeding up our awareness and enabling faster, and hopefully a better, understanding of what is going to come—the more aspects and the more dimensions of it, the better.

9 Quantum-Inspired Methodology and Its Revealed Prospects

A new wave of quantum-inspired methodology has quietly and yet sturdily gained momentum in all of the social sciences, including the humanities. Therefore, we here were merely picking up the ball where others have propelled it already (cf Hakemi et al., 2022; Wu et al., 2021; Surov et al., 2021; Wilson, 2016; Wendt, 2015; Oppermann, 2015; Busemeyer & Bruza, 2012; Coale, 2011; Filk and von Müller, 2009; Barad, 2007; Busemeyer et al., 2006; Beck, 2001; Zohar & Marshall, 1993; Carter, 1989).

First, the book was set to stride along old paths, the path of Friedrich Nietzsche, trying to describe and explore all aspects of *human nature* by way of philosophy (for the purpose of philosophy) (cf esp. Nietzsche, 1878), as well as the path of Niklas Luhmann (1984, 1997) trying to look for and understand human *inter-complexities* and their role in everyday life, and society as a whole, as well as all of its institutions (from marriage to government bureaucracy).

However, one discovery led to another, and a theory formed, and the insight that the initial temporary randomness (that was supposed to be fit into system of chapters) was needed, and revealed a great deal of creative potentials, as experiences about one quarter in on the way into this book have shown.

Retrospectively, it amazes me to see and realize the importance of initial appearances of key words, key thoughts and key themes in the first entries and first parts of this book, that later on, in the succeeding parts of the book, reformed my thought process, and refashioned and reinvented the very same. One just has to look for words like 'force', 'quantum', etc. early on in the beginning of this book, and thus see the workings of quantum-inspired methodology, the importance of relative randomness and parallelity, while ensure full compatibility (universality of observations enabled by their least common denominator, i.e. words, the atoms of communication). The strategy and method applied fully worked, one cannot hope for more than that.

Creativity, for sure, so is the experience of this book, and that of others, is a key advantage of here and there trying to set out on a quantum-inspired research work, or a serious of quantum-inspired research works (research program); the latter being perhaps more easily achieved, compared to doing it all in one study.

Doing things in parallel was from the beginning the practical thing to do, and the only way to do according to the newly envisioned and newly set up methodology of quantum research in social science and humanities research. As always, every own discipline may work and adapt and reinvent their own discipline-specific methods, methodological principles and specific methods at hand.

Randomness and parallelity, and having a least common denominator of each and every observation and research objects, was crucial and thus enabled and ensured respectively the universality of the research study, meaning, the universal coverage of, as much as possible, all kinds (not all) of research dimensions and research problematics, plus all a more-or-less (not fully) random integration of the infinite universe of research objects under scrutiny.

The methodology applied in this book, hence pays tribute, and has to (in terms of feasibility and practicability) to the idea of *fuzzy logic* (for the importance of fuzziness in quantum mechanics, cf Granik & Caulfield, 2001; Schmitt et al., 2009; Gentili, 2018), where more, and much more will *suffice*; and less, and much less is better in terms of the conventional methods and their principles and ways of doing it along the old and trodden paths is hindering rather than helping. Fuzzy qualities/quantities of randomness and parallelity are cornerstones of this new method under this quantum-inspired methodological paradigm.

New things come from new things. With new methods, come new breakthroughs. This is not the end of the road, but opposite, it is the promising beginning of a new road with more road forks further ahead. The race is on.

10 The Sky Is Not the Limit—There Is No Limit

To provide a conclusion of conclusions provided, I think there is no better way do to that, other than *pointing out to the sky*, and say, *that is 'not' the limit*, we will travel much farther, and we will leave it to others to yet travel much farther, out into the unknown.

This is what sailors and scientists do. They travel out, they venture out into new territories, and bring back a new understanding, a sharper and more meaningful understanding of not what is going on just out there, but in us, in our worlds of communicative systems that we have created, and are creating—and that also create us, with every word we utter, with every thought we produce, in as it seems ever more pronounced and intertwined, and ever faster, ways.

11 The Strong Cold Winds Ahead

Unfathomable dangers are looming ahead of us. They are all brought upon us by ourselves: the shift in climate, the pollution of the earth, inhumane levels of inequality and poverty, vicious cycles of exploitation, oppression, discrimination, hate, violence and wars, and ever new technologies that threaten our very own existence and well-being.

We need to learn what made us do them, and not prevent them, and not undo them. We need to find cures to our own doings, the ways of doing things and not doing things. We need to free ourselves from blindness and then from the binds of our entanglements, and from the binds that form our entanglements. New, benevolent entanglements are to take their place.

Thus, *decoherence* (de-entanglement) from a bad entanglement is a good thing, and must be 'encouraged' (i.e. established and/or amplified). At the same time, *coherence* to good entanglements must also be established and/or amplified (cf Crull, 2021).

This sounds more difficult than it is, if devised and implemented well (see the effects of e.g. Advertisements in the Public Interests, APIs, cf Aspalter, 2014, 2015; HKISD, 2023). However, the worse the problem, the more efforts are needed for its cure. These efforts then also shall bear the greatest levels of effectiveness (the more prompt the better, the more lasting the better) and efficiency (output to input ratio). An energetic attitude, great compassion and an open mind will be quintessential on the way forward.

An infinite number of hearts and souls of today and the future need to find a solution. They are looking for a solution, one at a time, one after the other—one de-entanglement after another, one new entanglement after the other. It would be great if we choose to be progressing without regressing, advancing towards decency and compassion, away from greed and violence.

Solutions are there, every step of the way, either already in practice in rare form or outright perceivable (cf Aristotle, cited in Bauböck, 2008). Practice-based, comparison-based normative science (e.g. Aspalter, 2023a), strong logic and a strong tradition of critical thinking will show the way. Minding our steps, minding our ways will help to identify and make new choices, less distorted and more benevolent choices.

Science and wisdom create understanding. It is never too late, but looking out there, one can see it is getting cold out there, really cold. The cold winds have arrived already and they carry warnings of the future.

Epilogue

This book has been new to the author, in terms of writing and planning it. First, there are no chapters. This book is like 'one' theater act, one 'noodle of communication itself', as the theory of this book has described human communication in general, but also that of cultures, languages, countries, communities and in-between and inside people, and—logically, too—the communication system of any book.

To allow for more randomness, more switching, the chapters that usually structure a book are not there, as the methodology demands switching, freedom to maneuver, reaction to follow reaction, interaction to be (in a non-finite manner) multidimensional and parallel at the same time.

The titles for each part of the book are to be seen merely as 'sticky notes', which are only meant to shortly depict in what stage of the investigation we are in. They were added a posteriori, just to give a hint what mainly was going on in the book, and how the book was proceeding forward. It is *not* like an imaginary ancient Greek chorus (symbolizing the Gods) would come and sing about what is about to happen (a new chapter's content), putting their will onto the course of action so to speak. These are just sticky notes—for a posteriori purposes (for reflection, readers' guidance, etc.)—that have been assigned to each part (smaller, more digestible part) of the book.

In this way, this has been enabling, and is now highlighting, the *free realm* of the investigation itself, which was made possible by way of aiming at and allowing for *free and frequent* switching of topics, focuses, investigation levels and perspectives, as well as particular on-the-spot purposes in the development of the investigation.

Still, it is one investigator chasing *one goal*, the formation of *a new universal theory of human communication (and action) and with it the testing of a new methodology* (not just a method, but a new arena of methods). The new methodology is, and has set itself to be, *inspired by quantum mechanics* itself, as well as and particularly also *quantum computing.*

The turn of events, the turn of focuses by and large, and in detail, were not and are not supposed to be known in the beginning of any book or study that uses a *quantum-inspired method.* Freedom and creativity, and necessity to be able and

C. Aspalter, *Human Entanglement Theory*, https://doi.org/10.1007/978-981-97-5161-7

properly equipped to chase the goal (suspect/culprit so to speak), demand nothing less than that.

The *free realm and high reaction capacity* of the investigator are mandatory for success, or better, high likelihood of a success, of the investigation. Any *crime scene investigator* must be able to freely chase the suspect—i.e. the research goal, in the case of a quantum-inspired method, as this is what we applied and developed over the course of this book.

The investigator has to keep an open mind, follow the lead (and not a pre-determined script) and an eye out for any other not yet considered suspect or group of suspects, turn of events, and for replacing or adapting the/any lead theory once it has been picked up, doing all this along the way, building theory while investigating.

Hence, here, we have been applying the general *abductive method* of theory development in science. The quantum method (quantumology) is a particular—and new—application of an abductive method, in the realm of *abductive methodology*, following more closely (but not identically) also what Charles Saunders Pierce (1878a, 1878b, 1893) had in mind over the years. Modern interpretations of abduction have chosen to be too particular in their meaning and miss the point in terms of its general and much wider (than most people first imagine) application, as most of science is using this method already without realizing and acknowledging it.

For the experienced and more sensible scientific eye, it is clear that, research programs and series of research works on the same investigated research question/topic/purpose are the common ground on what scientific developments and breakthroughs are standing on today, in all of the world and in all disciplines—and *not* tiny and fully insulated and independent research works (which are, however, omnipresent, and rightfully so, when it comes to writing students' theses).

For this, one has to apply a wholesome perspective and look at breakthroughs and how theories (i.e. empirical theories)—the core of scientific knowledge, the bread and butter of scientific knowledge—have been and/or are being created, by the very most of us in science.

With regard to the newly-perceived, and now tested (its usefulness, and the applicability of its new specific methodological principles), method of quantum-inspired research (quantumology), over the course of an ongoing research investigation, the investigator depending on his/her particular field may have their own experience to bring to the table that is indeed very much needed for this or that research field, research problem and topic. The same is true with a crime scene investigator. Everyone needs experience, learning from past failures and successes, learning from everyone's failures and successes. Bringing in the knowledge of context, and first-hand knowledge about the research objects and/or problems, as well as of course knowledge about competing and other conventional theories, their limitations and strengths, and their mistakes and failures is required.

Pertaining to the all-important principles of this newly-perceived quantum-inspired scientific methodology, the here applied method is in fact only *as-much-as-possible(!)* random, parallel and all-encompassing/'universal' (in terms of coverage of different or different kinds of research objects/observations)—as we deal with an infinite number (and infinite types) of research objects, i.e. human entanglements that derive from and are based on all sorts of human communication. The second-order principle, here, is to *maximize* these first-order principles (as-much-as-possible randomness, as-much-as-needed parallelity, plus full-fledged universality).

These very same principles apply to *all cases* with infinite number of research objects, as well as a finite number of research objects (e.g. elderly people in China, in provinces, or cities of China).

The problem is just that we either deal with an *infinite*, or a too high (as regards the research means available), or a very high, number of suspects, research objects under investigations. In our case here, it is an infinite number of types and absolute human entanglements. Same with quantum computing, where there is a *will, strategy and a functioning method*, there is a solution. The human element of research investigation is wanted. It is also present in any code written for any computer, classical or quantum computer. This needs to be understood, acknowledged and appreciated, and thus built into theory-making of science and philosophy, as well as paid attention to in the making of theories in the applying sciences themselves.

Working with what one has is vital, like with the investigator out in the field. Where 'absolute randomness' is not possible (as it is still not possible with any computer nowadays) one works with as-much-as-possible levels of *randomness. Parallelity* is a must, but not in every instant, not on every page or in every smaller (smallest) section, and not to a too excessive degree (which cannot or could not be handled by the research investigator, research technique and/or setup, as well as research resources, such as time, money and/or publication space). But in general, by and large, parallelity is wanted and applied as a means to increase the creativity and power—and hence validity and usefulness—of the investigation.

In addition, *universality* of the subjects under investigation, that is their *least common denominators*, in our case this is *the unit of human communication*, is a precondition for this to work; just like in quantum computing there must be universal quantum gates, to facilitate proper calculation/proper logic and hence proper meaning of it all.

These are the core characteristics and the *founding principles* of the new method of quantum-inspired research conducted in this *philosophy-cum-social-science book*, which used and worked in philosophy to arrive at a new social science theory and methodology.

This book covers a lot of ground, philosophically, theoretically and methodologically. It is perhaps best consumed in small doses; or, for some, following an all-in immersion technique.

This book is certainly a product of the past, and certainly written for the future. The past and the future are infinite. The present is merely the switching from one to the other. What it brings, the future, is difficult and/or impossible to determine, especially the further into the future one wonders, wants or needs to look. This book was interesting (fun), and shall it prove to the useful in the future to a wider spectrum of applications and a wider audience of researchers, then I would have achieved my goal *to work fruitfully and to be useful*, and I would know that T.B. Bottomore and Karl Popper would agree! With regard to the application of concrete methodological principles, also both Immanuel Kant and Karl Popper, for sure, would have a lot to say. I would like to hear. And if Charles Saunders Pierce could join in the dialogue (and dialectic) that would be just great. All of us should join this discourse!

It is (as Popper also clearly pointed out) all about a competition of thoughts, of practicality of methodological scientific principles, of the formation of fruitful theories and, if possible, their successful empirical testing, logical and empirical corroboration (providing higher/highest possible degrees of confirmation) and hence foundation. If possible, theories, i.e. empirical theories, must be falsifiable (i.e. testable) or fallible. There are two kinds of theories, falsifiable ones and non-falsifiable ones.

No one can test string theory, in order to corroborate it or to bring it to fall, not now! So, according to Popper, this would not be a scientific theory, a conclusion which I refute, and any physicist, too, I think is inclined to refute. I sincerely think Popper has overshot the problem of demarcation and turned the problem of demarcation itself into a problem (for science). Popper offended many disciplines and many of their lead theorists and their followers, by way of his insensitive and razor-sharp demarcation.

Our theory here is falsifiable, but many other scientific theories (of the other kind of scientific theories) are not. Like in psychology, the theories of Freud, Adler and Vygotsky are not falsifiable. In sociology, Marxist/Neomarxist functional and conflict-based (actor-based) theories, as well as other (Non-Marxist) functional and conflict theories, are *not falsifiable—but they are all fallible*!

Elizabeth Steiner's two razors will take care of it (cf Steiner, 1988). First, non-theories are selected out, just calling something a theory does not make it a theory, i.e. *Steiner's razor no. 1.* On top, Steiner maintained that the competition, in each discipline, must be leaving all but one winner, the rest of them are to be discarded, they are out! This is *Steiner's razor no. 2.* We just need to apply *Steiner's two razors*, and *Ockham's razor* (relative simplicity is to be preferred) (cf Keele, 2010).

As long as they are real theories (see Steiner, 1988), non-falsifiable theories are scientific, all of them. They are being used by scientists, and that successfully (Popper indeed also highlighted this fact) in order to *gain* knowledge, *test* knowledge and *develop* knowledge—i.e. other theories, plus their findings and applications.

If one comes to the empirical—and in the end lasting, hence 'infallible'—conclusion that *free will* explains all (and every bit) of poverty, in all of the world, for all individuals—then our above constructed theory has failed. If one comes to the conclusion that poverty is only in smaller (or very small) parts the outcome of personal free will, but rather the combined outcome of the community one is born in, one works and lives in, conditions of one's family, the richness of one's parents, the tax burden of the poor, the social security financing burden of the poor, devastating effects of divorce laws and divorce rulings, the lack of health prevention and free-basic health care, etc. (cf e.g. Aspalter, 2023a; Brady, 2009, 2019; Brady et al., 2009; Brady & Sosnaud, 2010; Brady & Burton, 2019), then the theory of human entanglements has a very good chance to prevail, with *all its possible applications*, in *all possible fields of application.* The future will tell.

The theory human entanglement in fully testable, it is a full-fledged theory of universal scope (universal application). So, it does fall within the demarcation of scientific theories *à la* Popper; but that is not so important, since all theories are scientific theories—also those of e.g., just to name a few, Isaac Newton, Albert Einstein, Sigmund Freud, Alfred Adler, Lev Vygotsky, Karl Marx, Antonio Gramsci, Adam Smith, Carl Menger, Simon Kuznets, Thomas Piketty—all of which have to join *the competition*, and not all of them make it, i.e. great many of them will be, and have to be, selected out and discarded (cf Steiner's two razors, Steiner, 1988; as well as Ockham's razor, Keele, 2010). By the way, Einstein (about relativity in the macroscopic world), Vygotsky, Gramsci, Smith and Piketty, in my opinion and to my experience, have made the cut, and are the creators of the still, or the now, all-leading (all-beating) theories in their respective fields *or* streams of theories.

The perspective I have applied in this book stems from quantum mechanics and quantum technology, it is also a perspective of research programs, and long-term research investigations, and not that of small-sized insulated research studies.

Here, I would like to thank my university (Beijing Normal University-Hong Kong Baptist University United International College, Zhuhai), which has for long now instilled in me the importance of transdisciplinary research in science. In the first two teaching years here, in 2006 and 2007, I was given the chance to teach a course on philosophy (I simply was told to do it). Retrospectively, I know it paid off, new interests have simmered and grown over the years.

Back then, especially Immanuel Kant (1781), for me, was the key, not only what he said in his elaborated work, but more also what was said in between. The understanding I gained from that time, led also to this book here (a personal learning entanglement of mine).

Pertaining to the entanglements of thoughts and ideas, the switching, or dialectic as it was also later emphasized by Hegel (1991, 2010, 2018), I have also discovered for myself back then with my thoughts on Kant (1871). But, Hegel as Popper (2012: 353) noted, messed it up, the development of the dialectic, and for most it was forgotten in relative terms. Pierce did not agree with Kant, and thought the least of Hegel's works, as did Schopenhauer (Pierce, 1878a, 1878b, 1893). Then, until much later, Pierce (1878a, 1878b, 1893) picked up on the idea and method of the dialectic. And, again, most people using Pierce's term of abduction today, understand it and use it differently from Pierce or myself (they use it only as 'inference to the best explanation'). This is not how the term abduction is used in this book, as here I stay closer to the thoughts and ideas of Pierce, and combine them with the earlier thoughts of mine on Kant.

Kant concluded, and Popper backed him up, that the method used in science needs to be thorough by following its principles, and following the principles of the method chosen is what makes it 'scientific' (one key aspect at least). In general, we need to put more emphasis on creating and developing methods. Hence, this book also greatly, as one can see, has focused on the development of quantum-inspired methodology in the social sciences.

Abduction is a series of research actions/studies, this can happen over a researcher's lifetime, over many years, or just in one research work (as demonstrated here). The same reasoning as given by Popper (as regard to his method the way he perceived it) is applied, but they are being repeated, thus applying a different methodology, that of abduction, which was (gravely) overlooked (or ignored) by Popper.

Also, for this book, the ideas of Pierce and Kant were indispensable to the setting up and fleshing out of principles in abductive methodology. Today, abductive methodology has been already relatively widely accepted across different science, ever since the works of Pierce at least. But, there is some more work to be done in this direction. I have been working on it for about half a dozen years already in my head, and on paper, and I will possibly bring all of it down to paper soon (I hope). The key aspects, and examples, of using the abductive method I have already put out in various books of mine, especially in the scientific field of social policy, over the past years.

Kant said pure deduction is not possible, and so is pure induction. Popper distinguished his deductivist-empiricist method from pure/radical deductivism (pure rationalism), plus he also strongly opposed pure/radical inductivism (pure empiricism) as a scientific method. Popper's deductivist-empiricist method contains elements from both deduction and induction; logical derivation and generalization (respectively).

All of Popper's efforts hinge upon the problem of demarcation, i.e. deciding the demarcation for the classification of theories and concrete methods applied into scientific ones and non-scientific ones (while also seeing the complexity of *differences* in statements and methods applied by each research study or when used to form and develop a theory, I may note that it is hard, if not self-refuting, to try to classify all theories, and all research studies one by one, and the generalizations made by Popper are themselves highly problematic, i.e. fallible).

The black-and-white kind of view applied by Popper was only applied by those actually which he attacked most, the natural and technical sciences. This is out of date, and has been for a while now, the worldview of Popper.

Other more subtle ways lead to better results and more general acceptance. One way is to classify academic disciplines as a whole, and thereafter identify any scientific nature of its theories and methods. For the classification of scientific disciplines one can be looking at:

(i) the presence of universal theories (even different kinds of prescriptive, explanatory and normative universal theories),
(ii) the presence of clear methodologies and especially also *discipline-specific* methods;
(iii) the presence of a very substantial (not marginal) *discipline-specific* body of knowledge (that is *not imported* or borrowed from other disciplines/sciences);
(iv) the export and application of theories, methods and concepts to other sciences;
(v) the body of knowledge that has to be from and has to be about the whole world (and not just about developed Western countries), and
(vi) the usability and fruitfulness (the applicability) of the theories and methods under scrutiny.

Many theories that are called theories are not theories at all, they are simple or abstract models, or just ideas (cf Steiner, 1988). The final cut for theories and methods to be called (logically and reasonably, not habitually) scientific or not, is their usefulness in creating more theories, more knowledge, and in their applicability, its width and frequency, and the therefrom resulting outcomes (i.e. usability and fruitfulness). This is a relatively simple and effective way—an Ockhamian way—of classifying science, versus not-so-much science ('in between cases', 'on the way cases'), versus non-science.

Again, to be clear, this classification of which discipline can be called a science, or what theory *as a consequence* can logically be called a scientific theory, is to be used as a (Weberian) ideal-type classification ('just about cases', 'mixed cases', 'about cases' are possible, and allowed for, i.e. they are wanted) (cf Weber, 2012; Aspalter, 2020a, 2020b). This is in contrast to Popper who used merely a simple dichotomic—black-and-white, in-and-out kind of—typology.

Popper's falsifiability criterion is *not a must, but a plus.* The demarcation line used by Popper (the falsifiability criterion) does in fact exclude many physics theories (much of theoretical physics, see string theory, white/gray hole theory, parallel universes, etc.) and their theoretical methods, it would exclude mathematics (pure rationality), and most of social sciences and humanities, because of what and the way 'science' is done there. Here, *the demarcation line* was not helping, but actually *was a problem* by itself, and *posed many problems* for the further development of philosophy, and the especially the theory of science (his fixation with a demarcation line, and thus to be an eternal judge of who and what is to be called a scientist or scientific, made Popper less a philosopher *who wishes to be just* and more a commoner *who likes to judge*, cf The wish to be just and the wish to be a judge, Nietzsche, 2014: 283–284).

Popper made things very complicated; he negated induction, but supported quasi-induction; he opposed pure deduction, but supported deductivist-empiricist theory and method (a synthesis of both deduction and induction elements/steps).

Ockham's razor needs hence, if one looks deep and long enough, also be applied to Popper, most of his theory of demarcation of science versus non-science, not all things he said on that topic. Many things are gold, but not all, and certainly his theory of science is *not the gold standard anymore* for most scientists that call themselves scientists and work as scientists, it should not be.

Here, I have taken the liberty to carry on the new ventures into the *abductive methodology*, which have formed over the years and decades in modern-day science, and on top of that also apply a new method, which also has been circulating already for some years (if one takes a look around in other disciplines, e.g. literature or psychology), i.e. quantum-inspired methodology and theory-building.

Elizabeth Steiner's book on theory building (1988) has also played a formative impact on my thoughts and my experience with *theory building*, which has become a very much reoccurring and perhaps the identity-forming character of my past scientific life so far, this past quarter a century.

A paper of mine that has been published in 2010, on philosophy in science, has also made its way deep into my intellectual understanding of the philosophy and theory of science. And here, it is the work of Albert Einstein that has ever since left a decisive impact on how I understand abduction, and Pierce.

'Dreaming' in science, i.e. wondering and contemplating, is important, it is not 'guessing' and not 'having an intuition' as Popper would say (Einstein called it 'inventing', which is a good word for it). I define 'dreaming' (profoundly wondering/contemplating) as the creation of *well-informed and as much as possible theoretically and empirically-founded knowledge.*

Erwin Schrödinger's elaboration on the process of theory/model building and knowledge development is also particularly helpful (Schrödinger, 1935a), as well as in addition Charles Saunders Pierce (1878a, 1878b, 1893), also James William (1970, 1981) and John Dewey (1910, 1922) are remarkably helpful, as their stance is on *pragmatism*, and they see the need to cope with reality in science, not with fiction and/or abstract intellectual contemplations.

> Schrödinger mentioned (1935a: 52) that gradually we achieve—over longer research processes—an increasingly more fitting picture, i.e. our thoughts (models, theories), to the empirical facts.
>
> In Nietzsche's words: "The specifically qualitative aspect ... of every chemical process, still appears to be a 'miracle', as does every 'locomotion'; no one has 'explained' the push. And how could we explain! We are operating only with things that do not exists [quantum physics nowadays is basically confirming this(!)]—with lines, surfaces, bodies, atoms, divisible times, divisible spaces. How is explanation to be at all possible when we first turn everything into a *picture*—our picture! ... we learn to describe ourselves more and more precisely as we describe things and their succession. ... a *continuum* faces us, from which we isolate a few pieces ... The suddenness with which many effects stand out misleads us; it is a suddenness only for us. There is an infinite number of processes that elude us in this second of suddenness" (*second emphasis added*, Nietzsche, 2009a: 113; cf also Heisenberg, 1927a, 1927b; Born & Jordan, 1925, 1930; as well as Clark, 2008).
>
> We need to see cause and effect as a *continuum*; we need to see 'the stream of the event' (Nietzsche, 2009a: 113). This continuum, over the years and in this book, I have referred to as *the noodle of human communication (and action)* (cf e.g. Aspalter, 2023b).

To be properly equipped, to be able to achieve this, one needs to look at science from a realistic point of view. For science to be able to operate in a practical (hence, successful) manner, at a closer look, once one steps back a bit, one can see that 'dreaming' relates to both a priori (deductive) and a posteriori (inductive) knowledge development—as researchers seldom work on their first piece of work within a given field, research program and research investigation. That is, theories are known already, and they have been contested and evaluated already, and empirical reality has been measured and compared to theories already, again and again.

So, when Albert Einstein was thinking and pondering about light, when he was setting up and pursuing his thought experiments, he was of course conducting the abductive method. And this needs to be kept in mind, the most brilliant breakthroughs in science are based on the method (and methodology) of abduction.

'Dreaming', i.e. *as much as possible theoretically and empirically-founded—knowledge* creation (i.e. thinking about/thinking out/imagining), comes first. The deductive work comes later, and inductive work (measuring and testing) comes at the very end of one and the same research process (cf Aspalter, 2010). Or, the researcher may switch in between them *again and again*, or visit these steps *in recurrent circles*, which is the most common research practice. Both are to be understood as (Weberian) ideal types of research processes/practices (Aspalter, 2020a, 2020b; Weber, 2012).

Einstein first ‘dreamed’ (‘thought out’/set up the theory), then he came up with the formula of $e = mc^2$ and the other parts of his relativity theory, and, thereafter, his theory was proven right with empirical testing (by Arthur Eddington when measuring the planet mercury, and then again by the development of the atomic bomb later on). This happened not on the same day, not in seven days, but over many years.

It is not a revolution this book here is trying to bring about, but simply a revelation of what has already been going on and is going on in all of science. But then again, of course, the book turns yet again the wheel of science, theory and methodology, to increase knowledge and understanding, as it should.

Christian Aspalter
Traunkirchen am Traunsee, Austria

Addenda

Notes on Related Issues

On Kuhn Versus Popper

There has been a battle going on over predominance in the realm of theories of science. Popper's theory of science (2012) has been, or is in fact being replaced by the dominance of Thomas Kuhn's work (1970a, 1970b, 1977, 1978, 2000) and his theory of normal science. Both are mutually exclusive, so one has to go (cf esp. also Lakatos & Musgrave, 1970). Only one can be declared the winner, along the lines of Elizabeth Steiner (1988) one must always prefer the better explanation, the better description, the more useful theory. That is what can be called *Steiner's razor no. 2*. Even when dealing with very similar and similarly performing theories, one theory *has to*—so Steiner's second razor—be discarded.

In the case of both Popper's and Kuhn's theory things are simple, they are mutually exclusive. No one can arguably support both (reasonably and logically that is). While Popper's theory of science is suggesting a linear progression of breakthroughs and overall development of science itself, Kuhn has a very different theory, one where sudden steps forward are common trait of the development of science across centuries and across all sciences. The stagnant periods in between the sudden steps forward are what Kuhn calls '*normal science*', where no (real/major) progress is made, but theorists and all scientists are (i) merely copying each other, and/or (ii) are content with the development they have made so far, and/or (iii) prohibit one another from stepping out of line, and create, develop or support new ideas and hence theories (and methods).

Popper (1957, 2007, 2012) did not investigate his theories empirically, but Kuhn (1970a, 1970b, 1977, 1978) did. Kuhn's take on the development of theories and science as whole has been quickly ascending in fame and in its usage, across a great number of sciences and academic fields. While Popper's theory still has a strong followership, it is dwindling and falling back relatively compared to Kuhn's rising popularity and central figure in the realm of philosophy of science.

C. Aspalter, *Human Entanglement Theory*, https://doi.org/10.1007/978-981-97-5161-7

Popper was slashing out on particular sciences, (in general) calling them non-scientific. Popper was trying to support the deductive method as used in theory-dominated psychology science, trying to oppose the other side, the empiricist-dominated psychology science (that focuses on brain waves measurements, experiments with rats, etc.). In the end, Popper managed to upset most social science and humanities researchers that use all kinds of theories, 'falsifiable' ones and 'not falsifiable' (and yet fallible) ones. The latter, fallible theories, are also subject to intense competition and selection process based on Steiner's two razors, as well as Ockham's razor. They have to conscribe to a general competition of which one is the fittest—for both which type of theory and which particular *individual theory*. Popper also attacked front on the technical and natural sciences that are strongly positivist and inductive-oriented sciences.

Kuhn has lashed out, too. But, in a much more subtle way. He does not call names (so to speak), he does not judge (as Popper does). Kuhn reports, he shows evidence, and generally shares his concern. Many scholars *are made to not be creative* due to the ways in which the system of science works. This refers particularly to (a) universities in particular and their promotion and hiring and evaluation schemes, but also (b) the publishing world (especially with the double so-called 'blinded'-but-never-blind reviews, as people know each other's work, e.g. also by looking at the references, etc.).

Referees like to shoot down others, often for a scientific reason, and more often than not for personal reasons (when not liking a theory, a methodology, a particular method, not liking the references used and not used; and, of course, when not liking the suspected author/s of this or that paper, or this or that book). This is making science not creative, younger researchers in particular are risking their careers (their family's income and thus their family's future) if they support new ideas, create new theories and use new, or less often used, and less-liked methods.

So, Kuhn in fact while pointing to an uncomfortable truth, is also assigning the role of a victim to those researchers who also see themselves and their colleagues as victims of the system, of how the system is 'to be played'. People have to publish or perish, and publishing is only likely or possible if old theories and old (conventional) authors are referenced *and* supported. Many younger and not so young (who still can remember their early years, or who simply are understanding the game, and how it is played) are supporting Kuhn's theory, as they have experienced the forces that created it. They are bound, they are dealing with negative entanglements (for this, see below).

No wonder that the feedback in general suggests that Kuhn either is winning or has won already (cf e.g. Worrall, 2002; Van Kersbergen and Vis, 2015; Lang, 2013; Layman & Rypel, 2023). I have to fully support this conclusion. This is a tremendously important conundrum in especially the social sciences and humanities, but also far and wide within the realms of other, technical and natural, sciences.

Popper's theory is descriptive and normative (evaluating/judging and prescriptive at the same time, while leaving no room for freedom/deviation from his method and concept/plan). Kuhn's theory is descriptive and explanatory, and, implicitly so, also normative. Popper's theory, hence, lacks any sort of explanatory capacity. Kuhn's theory is stronger, it has greater range of functions and more useful applications (rather than being highly judgmental, and robbing most scientists their 'sacred' status of being a scientist) (cf Popper, 2012; Kuhn, 1970a, 1970b).

Kuhn has also a visionary element, he offers hope in the long run, while giving a reality check for the here and now, which is not easy to accept either. For Kuhn, the hope to change everything is alive, and this may encourage many to stick with this theory and to help propagate the very same, by simply referring to and debating it. In the case of Popper, scientists that work on one aisle of science are played against the other across the demarcation aisle. One side of science is shamed and, on top, robbed of support and opportunities. On this side, scientists are shamed, indirectly and rather directly also in real life, by a number of ways: much lower salaries, much lower research funding, lack of ideational support, name calling in general, as well as name-calling of courses e.g. 'math for science' versus 'math for non-science', and much more.

So, with Popper, hope is either granted or killed (taken away), depending in which science or academic discipline one is working in, and on what kind of research project or program.

For the sake of parsimony, we can come to a (here, by now) already-known conclusion: Popper is out, Kuhn is in. Time has decided, and the people, i.e. the scientists who themselves are producing knowledge (science), have given their preference (vote of confidence) regarding Popper's versus Kuhn's theory of science.

Human Entanglement Theory Fully Supports and Explains Thomas Kuhn's Theory of Normal Science

Picking up from above, in the fashion of thought entanglement, I must proceed to another additional question that is of interest here. What is the relationship between human entanglement theory and the *Theory of Normal Science* of Thomas Kuhn (1970a, 1970b)?

Humans are not entirely free, they in much what they do or want to do are bound, by money, by their peers, by rules and regulations, by administrations, by policies, and such. This does not mean that there is no free will, it is just that is very difficult for the ordinary (median) citizen to identify, ignore or shed off those many impeding, obstructing and restraining interactive causal connections, i.e. inter-complexities as I have also called them in this book, or simply human entanglements. At the same time, there is a lack of positive, enabling entanglements, rewards, such as, e.g. pay raises, additional job security, research funding and publication opportunities that are missing for creative and innovative, most meaningful, most up-to-date research undertakings and entire research programs over the course of researchers.

There are a number of different types of entanglements that are at play when it comes to the formation and sustaining of 'normal science'. First, there are social-cum-economic entanglements that push (force, compel) scientists to quote and make reference to—and support, defend and champion(!)—the most commonly known and supported scientists and their most commonly known research works, as everyone wants to be able to pass the review process in both journal and book publications.

This is what also largely is to blame when it comes to the *Matthew effect* in science (Merton, 1968), where the very top researchers and their very top publications get almost all the citations and dominate all of science. Merton named this effect after a passage in the bible.

A namesake, but of a different origin, is the case of the *Patrick Matthew effect*. As Sutton (2022) and Sutton and Griffiths (2023) demonstrate, Charles Darwin who published decades later, took over all the credit and fame for the theory of natural selection already established by Matthew in 1931 (Matthew clearly explained it at least twice and named it "natural process of selection", which Charles Darwin 28 years later re-named "process of natural selection" ;on top, Matthew's articles were reasonably widely read and cited); Wallace who wrote about natural selection, too, was also noticed, but Matthew, the original inventor of this theory, got one could say completely ignored by science and history, because of his lack of influence and power in academia (cf Matthew, 1831, 1832; Darwin, 1859, 1871; Wallace, 1855).

Failure to do so ends in not getting promoted, not being able to afford more (or the same) for one's family, partner and children, failure to do so ends in not having a good or no career in science. Publish or perish is the motto. It is also, depending on *its attached conditions*, chiefly responsible for a number of both social entanglement forces (partnerships, family situations, peers) and economic entanglement forces (salary, promotion, job security, future career concerns).

Second, in also quite many cases, rather very frequently, researchers are staying within their comfort zone, they stick to the theories and methods, and the authors, they have learned and got to know at their university when they were still students or young researchers. Many are getting offered tenure positions, which helps in fact (I do not argue against tenure!) to lock in creativity and lock out pressure to learn and do new things—and, in the following, to support and encourage others to do new things. These forces here can be called personal entanglements and personal situation entanglements.

Third, the perhaps most yet named reason, for the limited progress, or rather non-progress, during times of normal science, where business as usual dictates daily, yearly and lifetime scientific outputs, is the review process. It is problematic and helpful at the same time. Its helpful function/effect lays in selecting out very poor papers and studies (journal articles or articles that go into books). The other—by the publishing world entirely overlooked—effect is that the best, and the most innovative papers are also selected out by middle of the road referees, or referees that are not up to speed or level or expertise required to understand, judge and comment on the work under consideration. This last third kind of entanglements are professional-cum-negative-competitive entanglements.

A helpful way to remedy this last particular situation can be, for example, by calculating that the number of partial citations (each citation divided by the number of authors) and net citations (excluding any kind of self-citations), *and* by demanding that the referees shall (or must have) a relatively equal or higher number of citations than the author/s, and the field and sub-field of the work and the expertise of the referee must match. At the same time, there must be a system to ensure competitive theorists and competing theory/methodology-supporting referees do not use this opportunity to shoot down other works of other theoretical and methodological outline (which is possible to be avoided, but almost never ensured/applied in a functioning way).

These forces have been talked about in probably many hundreds of papers, presentations and lectures. This is not new. New is, however that a new theory, a theory of universal human communication and communication-based action is giving now theoretical support, a full logical theory—the universal theory of human entanglement.

This is to show the way for future research paths and ventures; and not to come up with new arguments, new reasons *why* we have to deal with the problems caused by phenomenon of 'normal science'. All in all, I am glad this new theory can help to further logically and theoretically corroborate (give support to) the theory of Thomas Kuhn (1970a, 1970b, 1977, 1978).

I hope this helps, in some part, to encourage many others to join in the theoretical debate and, especially, get down to the business and follow the call of science, and start also with new series of empirical testing, including triangulation of methods, cases and research data.

Human Entanglement Theory and Social Policy Science

In order to not withhold, due to the on purpose differently conceived outline of this book and the entirely different focus on philosophy, plus a general/universal theory of human communication (and action), as well as the development of a new quantum-inspired social science methodology, I would like to add here some more notes on the *formational* developments, which brought me here to start and follow through with this book project.

I will in the following talk about both the particular entanglements of this book to my previous books and articles, and my earlier experiences, as well as the entanglements with which this book is influencing the science of social policy (yes, social policy is a science, it now fulfills all the six criteria for any science to be a full-fledged science, a newly-emerged science, not a big science, but a thriving and expanding science indeed, see epilogue above).

Everything one does is affected by past communication (thoughts and actions included), and so is the theory of human entanglement; when looking at its roots, the combination of a post-Luhmannian theory, a theory of interpersonal plus intrapersonal communication. In doing so, it becomes then clear that this thought and theory of mine (also) goes back to the year 2006, as then I also invented *the metaphor of the noodle of communication*, and hence *'the noodle of communication' theory*. I learned about Luhmann by my law and politics professor in the mid-1990s. The previous acquaintance of mine with theory (and practice) of Buddhist psychology as taught by Tich Nhat Hanh, was putting things in place, in my *Gedankenwelt* (thought world). The inside world and the outside world of a person, the conscious world and the subconscious world of the human brain included, are all consisting of the same elements, words and feelings (which also are words, or can be understood and expressed by them, or passed on with them). To stick with Luhmann's pure social world was not enough, so early on I extended the reach of Luhmann's theory of social systems (or theory of social communication) to also include intra-personal communication, all types of intra-personal communication.

With the addition of Foucault's theory of power relations, and later on the addition of Nietzsche's world (ways) of looking at history, literature, language and hence culture as the source of hundreds of causations for all of our actions today, the main body of thought for some of my more recent works and of course this book here was formed.

The *Theory of Super Inequality*, and the *Theory of Distorted Choices*, from one of my most recent books (2023b) are most obviously related theories, as these share the same original roots of thoughts and theories—from Niklas Luhmann, to Tich Nhat Hanh, to Michel Foucault and Friedrich Nietzsche, to the unification of all of them based on their smallest common denominator, communication.

My study of Hinduism and Buddhism, and particularly the teachings of Krishna and Buddha, in the Bhagavad Gita and Anapanasati Sutra, starting from my late teen years, also came back to the fore, in this book, helping it to become (I hope) more well-rounded, in terms of a more global representation of thought.

More than half of my life I have lived in Asia, particularly in China. When I came to Asia the first time, when I was 19, I fell in love with it, especially with Malaysia and its Indian culture, but also Chinese and Korean culture later on. This all streamed into this book, as I am made, and this book is made out of my past. In this way, I can, for example, also say I am entangled to Hermann Hesse's *Siddhartha* (1922), a book I first read back in school, and a couple more times later on. Thus, considering all things to be considered, I may also say I am entangled all the way back to the teachings and words of Krishna, Buddha, and Confucius.

Western philosophy, I started to bond with much later, in 2006. I—like anyone else—am a mere vessel of the ways of the world, its highways of thoughts that were printed into books, and into the cultures and languages of the people over thousands of years, and those cultures, languages, stories/teachings and thought systems again are entangled with yet much earlier cultures, civilizations, peoples, and their ways of thinking and talking. This I see now.

To the question of what this has to do, how this is entangled with social policy science, it is like this: A year ago, I was teaching a course on social security in aging societies, as so often, and I started to give a greater emphasis on how old-age-income-maintenance systems, health care systems and long-term care systems are *all related to each other and all other social policies*. I said in this course often, *everything is related to everything*.

This means, and this has been already for long my philosophy of social policy (since back in the day when I was teaching at Seoul National University), that one cannot know social policy, if one only does focus on and learn from, and write about, one subfield of social policy, of which there are so many. Education and jobs, health care and poverty, inequality and poverty, housing and wealth, so many things are intrinsically connected to one another, they are causally entangled, and all policies in each field are causally entangled with other policies and non-policies of all other subfields. By doing one thing, examining one thing, putting out a policy and its measures for one thing, it affects all other things.

The *principle of functional equivalence*, furthermore, makes all sub-disciplines, their policies and non-policies, their systems and programs, their often poor design, and the lack thereof, even more entangled.

The generalist social policy scientist today needs to be in addition also at least a multiple expert social policy scientist, and the expert social policy scientist must also be not only read into, but also standing on firm ground in most other, or at least a number of other, subfields of social policy (education policy, housing policy, family policy, gender and sexuality policy, social security policy, anti-poverty policy, employment policy, regional development policy, health and mental health policy, taxation policy, economic policy, and so forth).

Also, what I have already been teaching for many years, in the classroom and in my books, when one talks about one country, and learns from one country, or two more, one knows (really knows) nothing about what to do in social policy, *what is good and why, what is not good and not so good and why.*

Explanations, evaluations and recommendations of how things affect everything else—how they are entangled (causally connected)—are *only possible through* intense and extensive, and repeated, and that as much as possible, *comparisons*, and *through tested and contested theories* that are practical, that come from the empirical world and are meant to be applied in the empirical world (not just in intellectual circles, academic forums and classrooms), and usually highly testable (falsifiable) empirically, when one just looks at *all* the cases studies and data already available from *around the world*.

All my social policy life, I have been concentrating on these two things: *First, comparisons*, learning from and based on as many countries as possible, also including their histories, politics, cultures and developments. *Second, building theories*, as there are not too many. I mean we have quite a lot of theories, real theories *à la* Elizabeth Steiner, also different kinds of universal theories, and lots of competing *own* theories in social policy science, *when compared to other, neighboring disciplines*. But, for me, they are never enough. The field of social policy is so big, and expanding fast (see climate change, demographic change, gender change, etc.), and so diverse in applications, and also so diverse with regard to types of policies and themes, types of social and societal problems. Hence, I have built or extended-cum-built quite a few theories in the past myself, to contribute to and to enable social policy science and hence social policy practice, and this universally, in all parts and corners of the world.

It is mandatory to grow science, and to grow science is indispensable to change politics, and to help make this world a better place, a better planet for people to live in fully decent conditions; without poverty and unnecessary extreme inequality that causes it; without mass hunger; mass diseases and debilitating health conditions; mass incarceration and mass violence; without physical wars; without economic, mental and cultural wars.

Like with my recent *Theory of Super Inequality* (Aspalter, 2022, 2023b), this theory here, on human entanglements in general, is also descriptive and explanatory, and not normative. One theory cannot fulfill all of these three functions—*to the fullest*—at the same time. There are other theories that are designed to be normative (evaluative and prescriptive), and this would be, for example, the *Developmental Social Policy* theory, first set up by James Midgley (1995, 1996, 1997, 2014, 2017) at the University of California at Berkeley and developed further by myself (2014, 2015, 2017, 2020, 2021, 2023a, as well as Midgley & Aspalter, 2017)—adding new policy methods and solutions, plus explaining them, and explaining why not to repeat *mistakes of the past that are so stubborn, so widely-copied around the world* (mostly also due to the path dependencies, i.e. Colonial entanglements, that have been set in motion a long time ago by former Colonization of much of the world by the West plus Japan) (cf e.g. Midgley & Piachaud, 2011).

These mistakes, these horrible choices (*distorted choices*), are *so cemented into people heads and books* (*entanglements*), that it is very difficult to even thematize them (due to yet other entanglements), not to mention to get rid of them and replace them with positive social policy choices, i.e. solutions with no bad side-effects and no bad follow-on effects, fully effective and fully efficient (and fully decent) social policy solutions (all due to different kinds of entanglements).

Here again, I can see nothing but human entanglements everywhere, all sorts of entanglements, more obstructing than enabling ones (so far). The *Human Entanglement Theory* is designed to do exactly that to help explain universally present shortcomings in human and social development, and shortcomings in all of the realms of social policy, most of them present in the very most parts of the world.

Notes To Go

The Capture of the Soul

Trick or treat,
lie and cheat,
with ev'ry tweet.
Bittersweet,
here, they bleed
their soul and wit.
Sweet lullaby,
Say goodbye.
You'r part of it,
you'r in the feed
and on the—grid.

The Policy Labyrinth

Security,
market,
flexicurity,
choice.
Equalizing,
shaming,
quieting,
cancelling.
Wrongdoing,
all the same.

One Theme, One Party at a Time

Who likes to go over
and greet the forces
that rule science and

C. Aspalter, *Human Entanglement Theory*, https://doi.org/10.1007/978-981-97-5161-7

communication?
Fools of all nations,
join in celebrations!
New buzzwords, and new themes,
for sure, it's hip and cool.
There, there is no need, and
no one asks, for effect,
and verification!

References and Further Readings

Alexander, M. (2020). *The new Jim Crow: Mass incarceration in the age of colorblindness.* The New Press.

Altman, I. (1975). *The environment and social behavior*. Brooks/Cole.

Altman, I. (1976). Environmental psychology and social psychology. *Personality and Social Psychology Bulletin, 2*, 96–113.

Ariely, D. (2009a). *Predictably irrational: The hidden forces that shape our decisions.* Harper.

Ariely, D. (2009b). The end of rational economics, *Harvard Business Review*, July–August. https://hbr.org/2009/07.

Ariely, D. (2010). *The upside of irrationality: The unexpected benefits of defying logic at work and at home.* Harper.

Aristotle. (2021). *Aristotle's posterior analytics.* translated by H.G. Apostle. Thomas More College Press.

Arndt, M., Zeilinger, A., & Hornberger, K. (2005). *Probing the limits of the quantum world.* https://physicsworld.com.

Aspalter, C. (2010). In Scientia Veritas et Spero. *Journal of Comparative Social Welfare, 26*(2/3), 107–116.

Aspalter, C. (2014/2016). *Social work services and developmental social policy.* In C. Aspalter (Ed.), *Social work in East Asia.* Ashgate/Routledge.

Aspalter, C. (2015). New perspectives for active ageing: The normative approach of developmental social policy. In A. Walker, & C. Aspalter (Eds.), *Active ageing in Asia.* Routledge.

Aspalter, C. (2017). Saving lives through health care policy: A developmental social policy perspective. In C. Aspalter, R. Gauld, & K. Teguh-Pribadi (Eds.), *Health care systems in developing countries in Asia.* Routledge.

Aspalter, C. (2020a). *Third modernity.* http://www.ssrn.com.

Aspalter, C. (2020b). Healthy aging. In D. Gu, & M. E. Dupre (Eds.), *Encyclopedia of gerontology and population aging.* Springer.

Aspalter, C. (2021). *Developmental social policy and active aging with high quality of life.* In F. Rojo-Pérez, & G. Fernández-Mayoralas (Eds.), *Handbook of active aging and quality of life: From concepts to* applications. Spanish National Research Council, Government of Spain. Springer.

Aspalter, C. (2022). Super inequality: A general theory of mass poverty. *Social Development Issues, 44*(2), 1–19.

Aspalter, C. (2023a). *Ten worlds of welfare capitalism.* Springer Nature.

Aspalter, C. (2023b). *Super inequality: Theoretical essays in economics and social policy.* Springer Nature.

Aspalter, C. (2023c) *Quantum society: The rise of fourth modernity.* http://www.ssrn.com.

Aspelmeyer, M., & Zeilinger, A. (2008). *A quantum renaissance.* https://physicsworld.com.

C. Aspalter, *Human Entanglement Theory*, https://doi.org/10.1007/978-981-97-5161-7

Baclawski, K. (2018). *The observer effect*. https://www.researchgate.net/publication/326795653_The_Observer_Effect.

Bales, K. (1999). *Disposable people: New slavery in the global economy*. University of California Press.

Barad, K. (2007). *Meeting the universe halfway: Quantum physics and the entanglement of matter and meaning*. Duke University Press.

Bauböck, R. (2008). Normative political theory and empirical research. In D. Porta, & M. Keating (Eds.), *Approaches and methodologies in the social sciences*. Cambridge University Press.

Bauman, Z. (1991). *Modernity and the holocaust*. Polity.

Bauman, Z. (2000). *Liquid modernity*. Polity.

Bauman, Z. (2003). *Wasted lives: Modernity and its outcasts*. Polity.

Bauman, Z. (2005). *Liquid life*. Polity.

Bauman, Z. (2007a). *Liquid times: Living in an age of uncertainty*. Polity.

Bauman, Z. (2007b) *Consuming life*. Polity.

Bauman, Z. (2013). *Liquid love: On the frailty of human bonds*. Polity.

Beck, F. (2001). Quantum brain dynamics and consciousness. In P. van Loocke (Ed.), *The physical nature of consciousness*. John Benjamins.

Beck, U. (1986). *Risikogesellschaft: Auf dem Weg in eine andere Moderne*. Suhrkamp.

Beck, U. (1997). *The reinvention of politics*. Polity.

Beck, U. (1999). *World risk society*. Blackwell.

Beck, U. (2000). *Brave new world of work*. Polity.

Beck, U. (2005). *Power in the global age: A new global political economy*. Polity.

Beck, U. (2006). *The cosmopolitan vision*. Polity.

Beck, U. (2007). Beyond class and nation: Reframing social inequalities in a globalized world. *British Journal of Sociology, 58*(4), 680–705.

Beck, U. (2009a). *World at risk*. Polity.

Beck, U. (2009b). *An interview with Ulrich Beck*. Max Planck Institute. https://www.mmg.mpg.de.

Beck, U. (2014). We do not live in an age of Cosmopolitanism but in an Age of Cosmopolitization: The 'Global Other' is in our midst. In U. Beck (Ed.), *Ulrich Beck: Pioneer in cosmopolitan sociology and risk society*. Springer.

Beck, U. (2016). *The metamorphosis of the world*. Polity.

Beck, U., & Grande, E. (2007). *Cosmopolitan Europe*. Polity.

Beck, U., & Sznaider, N. (2006). Unpacking cosmopolitanism for the social sciences: A research agenda. *British Journal of Sociology, 57*(1), 1–23.

Beck, U., & Sznaider, N. (2011). The self-limitation of modernity: The theory of reflexive taboos. *Theory and Society, 40*(4), 417–436.

Born, M., & Jordan, P. (1925). Zur Quantenmechanik. *Zeitschrift Für Physik, 34*, 858–888.

Born, M., & Jordan, P. (1930). *Elementare Quantenmechanik*. Springer.

Born, M., Born, H., & Einstein, A. (1971). *The Born-Einstein letters: Correspondence between Albert Einstein and Max and Hedwig Born from 1916 to 1955, with Commentaries by Max Born*. Macmillan.

Bottomore, T. B. (1972). *Sociology: A guide to problems and literature*. Vintage.

Brady, D. (2009). *Rich democracies, poor people*. Oxford University Press.

Brady, D. (2019). Theories of the causes of poverty. *Annual Review of Sociology, 45*, 155–175.

Brady, D., & Burton, L. M. (Eds.). (2019). *The Oxford handbook of the social science of poverty*. Oxford University Press.

Brady, D., & Sosnaud, B. (2010). The politics of economic inequality. In K. Leicht, & C. Jenkins (Eds.), *Handbook of politics: State and civil society in global perspective*. Springer.

Brady, D., Fullerton, A.S., & Moren Cross, J. (2009). Putting poverty in political context: A multi-level analysis of adult poverty across 18 affluent western democracies. *Social Forces, 88*, 271–300.

Buchanan, J. M. (2009). Politics without Romance: A sketch of positive public choice theory and its normative implications. In J. M. Buchanan, & R. D. Tollison (Eds.), *Public choice theory* (Vol. 2). University of Michigan Press.
Buchanan, J. M., Tullock, G., & Tollison, R. (1980). *Toward a theory of the rent-seeking society*. Texas A&M University Press.
Buettner, D. (2023). *Live to 100: Secrets of the Blue Zones*. https://web.staging.ftl.netflix.com/ng/title/81214929.
Buettner, D., & Skemp, S. (2016). Blue zones: Lessons from the world's longest lived. *American Journal of Lifestyle Medicine, 10*(5), 318–321.
Busemeyer, J. R., & Bruza, P. D. (2012). *Quantum models of cognition and decision*. University Press.
Busemeyer, J. R., Wang, Z., & Townsend, J. T. (2006). Quantum dynamics of human decision making. *Journal of Mathematical Psychology, 50*, 220–241.
Capellmann, H. (2021). Space time in quantum theory. *Foundations of Physics, 51*, 44. https://doi.org/10.1007/s10701-021-00441-0
Carter, S. (1989). "A Place to Step Further": Jack Spicer's quantum poetics. In F. Amrine (Ed.), *Literature and science as modes of expression*. Kluwer Academic.
Chapman, J. (1981). *The Vinca culture of South-East Europe: Studies in chronology, economy and society*. BAR Publishing.
Clark, A. (2023). *The experience machine: How our minds predict and shape reality*. Pantheon.
Clark, A. (2015). *Surfing uncertainty: Prediction, action, and the embodied mind*. Oxford University Press.
Clark, A. (2008). *Supersizing the mind: Embodiment, action, and cognitive extension*. Oxford University Press.
Clark, A. (1996). *Being there: Putting brain, body, and world together again*. MIT Press.
Clavin, W. (2019). *Untangling quantum entanglement*. Caltech Magazine, October 11. https://magazine.caltech.edu.
Coale, S. C. (2011). Quantum flux and narrative flow: Don DeLillo's entanglements with quantum theory. *Papers on Language and Literature, 47*(3), 261–294.
Coenen, C., Grinbaum, A., Grunwald, A., Milburn, C., & Vermaas, P. (2022). Quantum technologies and society: Towards a different spin. *NanoEthics, 16*, 1–6.
Collins, A., & Hancock, G. (2014). *Gobekli Tepe: Genesis of the gods: The temple of the watchers and the discovery of Eden*. Bear and Company.
Conant, J. B. (1957). Robert Boyle s experiments in pneumatics. In J. B. Conant (Ed.), *Harvard case histories in experimental science* (Vol. 1). Harvard University Press.
Confucius. (1893). Confucian analects. In J. Legge (transl.), *The Chinese classics* (Vol. 1). https://en.wikisource.org/wiki/The_Chinese_Classics/Volume_1/Confucian_Analects.
Crull, E. M. (2021). Quantum decoherence. In E. Knox, & A. Wilson (Eds.), *The Routledge companion to philosophy of physics*. Routledge.
Darwin, C. R. (1859). *On the origin of species by means of natural selection, or the preservation of favoured races in the struggle for life*. John Murray.
Darwin, C. R. (1871). *The descent of man, and selection in relation to sex*. John Murray.
De, M. (2020). Hilbert space in quantum mechanics, IBM. https://community.ibm.com/community/user/ai-datascience/blogs/moloy-de1/2020/04/17/points-to-ponder.
Dewey, J. (1910). *How we think*. D.C. Heath.
Dewey, J. (1922). *Human nature and conduct: An introduction to social psychology*. H. Holt.
Dewey, J. (1927). *The public and its problems*. H. Holt.
Diessner, R. (2019). *Understanding the beauty appreciation trait: Empirical research on seeking beauty in all things*. Palgrave Macmillan.
DM, Dailymotion. (2011). *Chinese Toddler Wang Yue Tragic Death*, video. https://www.dailymotion.com/video/xlu9ov.
Einstein, A., Podolsky, B., & Rosen, N. (1935). Can quantum-mechanical description of physical reality be considered complete? *Physical Review, 47*, 777–780.

Ertürk, K. A., & Whittle, J. (2015). Climate change, procrastination and asymmetric power. *World Economic Review, 2015*(5), 40–53.

Eubanks, V. (2018). *Automating inequality: How high-tech tools profile, police, and punish the poor*. St. Martins Press.

Ferrie, C. (2023). Quantum entanglement isn't all that spooky after all. *Scientific American*. https://www.scientificamerican.com.

Filk, T., & von Müller, A. (2009). Quantum physics and consciousness: The quest for a common conceptual foundation. *Mind and Matter, 7*(1), 59–79.

Fine, A. (1996). The shaky game: Einstein, realism and the quantum theory. University of Chicago Press.

FlorCruz, J. (2011). *China soul-searching after Toddler's death*, CNN. https://edition.cnn.com/2011/10/22/world/asia/china-toddler-reaction/index.html.

Foucault, M. (1954). *Maladie Mentale et Personnalité, Presses Universitaires de France.*

Foucault, M. (1962). *Maladie Mentale et Psychologie, Presses Universitaires de France.*

Foucault, M. (1963). *Naissance de la Clinique*. Presses Universitaires de France.

Foucault, M. (1969). *L'Archéologie du Savoir*. Gallimard.

Foucault, M. (1975). *Surveiller et Punir: Naissance de la Prison*. Èditions Gallimard.

Foucault, M. (1976). *Histoire de la Sexualité: La Volonté de Savoir* (Vol. 1). Èditions Gallimard.

Foucault, M. (2010). *The government of self and others*. Palgrave Macmillan.

Foucault, M. (2014). *Wrong-doing and truth-telling: The function of avowal in justice*. Edited by F. Brion and B.E. Harcourt. University of Chicago Press.

Foucault, M. (2009). *Security, territory, population*. Edited by M. Senellart. Palgrave Macmillan.

Freud, S. (1921). *Massenpsychologie und Ich-Analyse*. Psychoanalytischer Verlag.

Galbraith, J. K. (1992). *The culture of contentment*. Houghton Mifflin.

Galbraith, J. K. (1996). *The good society: The Humane Agenda*. Houghton Mifflin.

Galbraith, J. K. (1999) [1958]. *The affluent society*. Penguin.

Galbraith, J. K. (2009) [1954]. The Great Crash. Penguin.

Gentili, P. L. (2018). The fuzziness of the molecular world and its perspectives. *Molecules, 23*(8), 2074.

Gil, D. G. (1996). Preventing violence in a structurally violent society: Mission impossible. *American Journal of Orthopsychiatry, 66*(1), 77–84.

Gimbutas, M. (1973). Old Europe c. 7000–3500 B.C.: The earliest European civilization before the infiltration of the Indo-European peoples. *Journal of Indo-European Studies, 1*, 1–20.

Gimbutas, M. (2001). *The living goddesses* (ed. by M.R. Dexter). University of California Press.

Giroux, H. A. (2010). Zombie politics and other late modern monstrosities in the age of disposability. *Policy Futures in Education, 8*(1). https://journals.sagepub.com.

Giroux, H. A. (2021). *Zombie politics and culture in the age of casino capitalism*. Peter Lang.

Gramsci, A. (2011a). *Prison notebooks* (Vol. 1). Columbia University Press.

Gramsci, A. (2011b). *Prison notebooks* (Vol. 2). Columbia University Press.

Gramsci, A. (2011c). *Prison notebooks* (Vol. 3). Columbia University Press.

Granik, A., Caulfield, H. J. (2001). Fuzziness in quantum mechanics. *Physics Essays, 9*(3). https://doi.org/10.4006/1.3029260.

Greenbaum, D., & Gerstein, M. (2023). Calculating the future: A quantum computing primer offers insights into the technology's most promising potential applications. *Science, 380*(6645), 589.

Greenfield, S. (2002). *The private life of the brain*. Penguin.

GWR, Guinness World Records. (2023). Website. https://www.guinnessworldrecords.com/world-records/first-recorded-strike.

Habermas, J. (1981a). *Theorie des kommunikativen Handelns: Handlungsrationalität und gesellschaftliche Rationalisierung* (Vol. 1). Suhrkamp.

Habermas, J. (1981b). *Theorie des kommunikativen Handelns: Zur Kritik der funktionalistischen Vernunft* (Vol. 2). Suhrkamp.

Habermas, J. (1987). Excursus on Luhmann's appropriation of the philosophy of the subject through systems theory. In *The philosophical discourse of modernity: Twelve lectures*. MIT Press.

Habermas, J. (1988). *Der philosophische Diskurs der Moderne: Zwölf Vorlesungen.* Suhrkamp.

Habermas, J. (1994). *Moderne und postmoderne Architektur.* In W. Welsch (Ed.), *Wege aus der Moderne.* Akademie Verlag.

Habermas, J. (2004). *Wahrheit und Rechtfertigung* (2nd ed.). Suhrkamp.

Habermas, J. (2005). *Die Zukunft der menschlichen Natur: Auf dem Weg zu einer liberalen Eugenik?* Suhrkamp.

Hakemi, S., Houshmand, M., KheirKhah, E., & Hosseini, Abed, S. (2022). A review of recent advances in quantum-inspired metaheuristics. In Evolutionary intelligence. https://doi.org/10.1007/s12065-022-00783-2.

Han, S.-J., & Shim, Y.-H. (2010). Redefining second modernity for East-Asia: A critical assessment. *British Journal of Sociology, 61*(3), 465–488.

Hanh, T. N. (1993). *Dharma talks*, given in July/August 1993, in Plum Village, Thénac, Dordogne, France.

Hanh, T. N. (2001). *Anger: Buddhist wisdom for cooling the flames.* Rider.

Hanh, T. N. (2007). *Living Buddha, Living Christ.* Riverhead Books.

Hanh, T. N. (2010). *Reconciliation: Healing the inner child.* Parallax Press.

Hanh, T. N. (2015a). *The heart of the Buddha's teaching: Transforming suffering into peace, joy, and liberation.* Harmony.

Hanh, T. N. (2015b). *No mud, no lotus: The art of transforming suffering.* Parallax Press.

Hayes, A. S. (2020). The behavioral economics of Pierre Bourdieu. *Sociological Theory, 38*(1). https://doi.org/10.1177/0735275120902170.

Hegel, G. W. F. (1991). *Encyclopaedia of philosophical sciences [Enzyklopädie der philosophischen Wissenschaften]*, translated by T.F. Geraets, W.A. Suchting and H.S. Harris, Hackett.

Hegel, G. W. F. (2010). *The science of logic [Wissenschaft der Logik], translated by George di Giovanni.* Cambridge University Press.

Hegel, G. W. F. (2018). *The phenomenology of spirit [Phänomenologie des Geistes], translated and edited by Terry Pinkard.* Cambridge University Press.

Heisenberg, W. (1927a). Über den anschaulichen Inhalt der quantentheoretischen Kinematik and Mechanik. *Zeitschrift Für Physik, 43*, 172–198.

Heisenberg, W. (1927b). Über die Grundprincipien der 'Quantenmechanik.' *Forschungen und Fortschritte, 3*, 83.

Heisenberg, W. (1979) [1927c]. Letter to Pauli, In W. Pauli (Ed.). Wissenschaftlicher Briefwechsel mit Bohr, Einstein, Heisenberg u.a. (Vol. 1): 1919–1929. Springer.

Herman, E. S., & Chomsky, N. (2002). *Manufacturing consent: The political economy of the mass media.* Pantheon.

Hesse, H. (1974) [1922]. *Siddhartha: Eine indische Dichtung.* Suhrkamp.

HKISD, Hong Kong Government, Information Services Department. (2023). Website. https://www.isd.gov.hk/eng/api_more.htm.

Holcombe, R. G. (2018). *Political capitalism: How political influence is made and maintained.* Cambridge University Press.

Holcombe, R. G. (2021). Elite influence on general political preferences. *Journal of Government and Economics.* https://doi.org/10.1016/j.jge.2021.100021.

Holcombe, R. G. (2022). Capitalism, cronyism, and inequality. In G. P. Manish, & S. C. Miller (Eds.), *Capitalism and inequality: The role of state and market.* Routledge.

Hou, R. Y. T. (2023). *Discussion with*, September, Zhongshan/Zhuhai.

Inoua, S. M., & Smith, V. L. (2019). *Classical economics: Lost and found.* www.researchgate.net.

Ives, P. (2004). *Language and hegemony in Gramsci.* Pluto Press.

Jablonka, E., Ginsburg, S., & Dor, D. (2012). The co-evolution of language and emotions. *Philosophical Transactions of the Royal Society b: Biological Sciences, 367*(1599), 2152–2159.

Kant, I. (1996) [1781]. *The critique of pure reason.* Translated by W.S. Pluhar. Hackett.

Keele, R. (2010). *Ockham explained: From razor to rebellion.* Open Court Press.

Kemp, M. (2006). *Leonardo da Vinci: The marvellous works of nature and man.* Oxford University Press.

Keynes, J. M. (1936). *The general theory of employment, interest and money.* Macmillan.

Komarzyca, D., & Fras, J. (2020). Language and politics in India and China: A cross-cultural comparative study. *Polish Political Science Yearbook, 49*(4), 9–36.

König, R. (2022). Towards a unity of theoretical and practical reason: On the constitutive significance of the transcendental dialectic. *Open Philosophy, 5*(1), 622–635.

Kropotkin, P. A. (1902). *Mutual Aid: A factor of evolution.* McClure Phillips.

Krugman, P. (2020). *Arguing with zombies: Economics, politics, and the fight for a better future.* W.W. Norton.

Kuhn, T. S. (1970a) [1962]. *The structure of scientific revolutions.* University of Chicago Press.

Kuhn, T. S. (1970b). Logic of discovery or psychology of research? In I. Lakatos, & A. Musgrave (Eds.), *Criticism and the growth of knowledge.* Cambridge University Press.

Kuhn, T. S. (1977). *The essential tension: Selected studies in scientific tradition and change.* University of Chicago Press.

Kuhn, T. S. (1978). *Black-body theory and the quantum discontinuity.* Clarendon Press.

Kuhn, T. S. (2000). *The road since structure.* University of Chicago Press.

Kwok, E. S. T. (2006). *Conversation with,* Zhuhai.

Lakatos, I., & Musgrave, A. (Eds.). (1970). *Criticism and the growth of knowledge.* Cambridge University Press.

Lang, A. (2013). Discipline in crisis? The shifting paradigm of mass communication research. *Communication Theory, 23,* 10–24.

Lavoie, D., & Chamlee-Wright, E. (2002). *Culture and enterprise.* Routledge.

Layman, C. A., & Rypel, A. L. (2023). Beyond Kuhnian paradigms: Normal science and theory dependence in ecology. *Ecology and Evolution, 17*(3), e10255.

Le Bon, G. (1877). *Histoire des Origines et du Développement de l'Homme et des Sociétés,* Paris.

Le Bon, G. (1894). *Les Lois Psychologiques de l'Évolution des Peuples,* Éditions F. Alcan.

Le Bon, G. (1896) [1895]. *The crowd: A study of the popular mind* [*Psychologie des Foules*]. Macmillan.

Le Bon, G. (1918). *Hier et Demain: Pensées Brèves,* Éditions Flammarion.

Le Bon, G. (1988) [1881]. *L'Homme et les Sociétés—Leurs Origines et leur Histoire,* Les Éditions Jean-Michel Place.

Leibowitz, A. H. (1974). Language as a means-of social control: The United States experience, paper presented at the *Annual Meeting of the World Congress of Sociology,* Toronto, August 8.

Leydesdorff, L. (2000). Luhmann, Habermas, and the theory of communication. *Systems Research and Behavioral Science, 17*(3), 273–288.

Li, Z., Ding, Z., & Zhao, P. (2023). Comparison of functional disabilities, place of death and end-of-life medical expenditures among centenarians and non-centenarians in China: A series of cross-sectional studies. *BMC Geriatrics, 23*(402). https://doi.org/10.1186/s12877-023-04111-w.

Longhi, P. (2023). *Mass incarceration: A graphic history.* Manic D Press.

Luhmann, N. (1984). *Soziale Systeme.* Suhrkamp.

Luhmann, N. (1997). *Die Gesellschaft der Gesellschaft.* Suhrkamp.

Lukes, S. (2005). *Power: A radical view.* Palgrave Macmillan.

Macdaniel, I. (2016). *The politics of historical economics: Wilhelm Roscher on democracy, socialism and caesarism.* Cambridge University Press.

Marcuse, H. (1955). *Eros and civilization: A philosophical inquiry into freud.* Beacon Press.

Marcuse, H. (1964). *One-dimensional man: studies in the ideology of advanced industrial society.* Beacon Press.

Marcuse, H. (1969). The relevance of reality. In *Proceedings and Addresses of the American Philosophical Association, 42,* 39–50.

Martin, J. (2022). *Hegemony.* Polity Press.

Mark, J. J. (2017). The first labor strike in history. *World History Encyclopedia.* https://www.worldhistory.org/article/1089/the-first-labor-strike-in-history.

Marx, W., & Bornmann, L. (2010). How accurately does Thomas Kuhn's model of paradigm change describe the transition from the static view of the universe to the big bang theory in cosmology? A historical reconstruction and citation analysis. *Scientometrics, 84*, 441–464.
Matthew, P. (1831). *On naval timber and arboriculture: With a critical note on authors who have recently treated the subject of planting*. Adam Black.
Matthew, P. (1832). Nature's law of selection. *Gardeners' Chronicle and Agricultural Gazette* (pp. 312–313).
McMurty, J. (2009). *Taxation and poverty: Injustice built into our tax system hurts poor the most*. Canadian Centre for Policy Alternatives. https://policyalternatives.ca.
Mead, G. H. (1925). The genesis of the self and social control. *International Journal of Ethics, 35*(3), 251–277.
Meagher, B. R. (2019). Ecologizing social psychology: The physical environment as a necessary constituent of social processes. *Personality and Social Psychology Review, 24*(1). https://doi.org/10.1177/1088868319845938.
Medinaceli Baldivieso, Agnes. (2017). *Procrastination and public policy*. https://inesad.edu.bo/developmentroast/2017/09/procrastination-and-public-policy.
Merton, R. K. (1968). The Matthew effect in science: The reward and communication systems of science are considered. *Science, 159*(3810), 56–63.
Midgley, J. (1995). *Social development: The developmental perspective in social welfare*. Sage.
Midgley, J. (1996). Toward a developmental model of social policy: Relevance of the third world experience. *Journal of Sociology and Social Welfare, 23*(1), 59–74.
Midgley, J. (1997). *Social welfare in global context*. Sage.
Midgley, J. (2014). *Social development: Theory and practice*. Sage.
Midgley, J. (2017). Social investment: Concept, uses and theoretical perspectives. In J. Midgley, E. Dahl, & A. Conley Wright (Eds.), *Social investment and social welfare: International and critical perspectives*. Edward Elgar.
Midgley, J., & Aspalter, C. (2017). Developmental social policy: Theory and implementation. In C. Aspalter, & K. Teguh-Pribadi (Eds.), *Development and social policy: The win-win strategies of developmental social policy*. Routledge.
Midgley, J., & Piachaud, D. (Eds.). (2011). *Colonialism and welfare: Social policy and the British imperial legacy*. Edward Elgar.
Milgram, S. (1963). Behavioral study of obedience. *Journal of Abnormal and Social Psychology, 67*(4), 371–378.
Milgram, S. (1974). *Obedience to authority: An experimental view*. Harpercollins.
Mises, L.v. (1944a). *Bureaucracy*. Yale University Press.
Mises, L.v. (1944b). *Omnipotent government: The rise of the total state and total war*. Yale University Press.
Mohan, B. (1992). *Global development: Post-material values and social praxis*. Praeger.
Mohan, B. (1993). *Eclipse of freedom: The world of oppression*. Praeger.
Mohan, B. (2007). *Fallacies of development: Crises of human and social development*. Atlantic.
Mohan, B. (2011). *Development, poverty of culture, and social policy*. Palgrave Macmillan.
Mohan, B. (2020). Ideologies of development. *Social Development Issues, 42*(2), 47–53.
MSN. (2023). *Father of 12 collapses and dies after foam begins 'pouring from his mouth'*, https://www.msn.com/en-us/news/world/father-of-12-collapses-and-dies-after-foam-begins-pouring-from-his-mouth/ar-AA1fWSYP.
Murphy, R. (1988). *Social closure: The theory of monopolization and exclusion*. Clarendon Press.
Myrdal, G. (1969). *Objectivity in social research*. Pantheon.
Nietzsche, F. W. (1887). *Zur Genealogie der Moral: Eine Streitschrift*. C.G. Naumann.
Nietzsche, F. W. (2008) [1883/1885]. *Thus spoke Zarathustra*. Cambridge University Press.
Nietzsche, F. W. (2009a) [1882/1887]. *The gay science*. Edited by B. Williams. Cambridge University Press.
Nietzsche, F. W. (2009b) [1888]. Ecco Homo. In A. Ridley, & J. Norman (Eds.), *The anti-christ, ecco homo, twilight of the idols and other writings*. Cambridge University Press.

Nietzsche, F. W. (2014) [1878]. *Human, all too human: A book for free spirits.* Wordsworth Editions.

Niskanen, W. A. Jr. (1968). The peculiar economics of bureaucracy. *The American Economic Review, 58*(2), 293–305.

Niskanen, W. A. Jr. (1971). *Bureaucracy & representative government.* Aldine-Atherton.

Niskanen, W. A. Jr. (1975). Bureaucrats and politicians. *The Journal of Law & Economics, 18*(3), 617–643.

Niskanen, W. A. Jr. (1994). *Bureaucracy and public economics.* Elgar.

Oppermann, S. (2015). Quantum physics and literature. *Anglia - Zeitschrift Für Englische Philologie, 133*(1), 87–104.

Orwell, G. (1949). *1984.* Books & Coffee.

Pauli, W. (1979). *Wissenschaftlicher Briefwechsel mit Bohr, Einstein, Heisenberg u.a.* (Vol. 1): 1919–1929. Springer.

Pearson, C. (2022). *Leon Battista Alberti: The Chameleon's eye.* Reaktion Books.

Peirce, C. S. (1878a). Deduction, induction, and hypothesis. In *Peirce, C. S. 1931–1958* (Vol. 2, §§ 619–644).

Peirce, C. S. (1878b). The probability of induction. In *Peirce, C. S. 1931–1958* (Vol. 2, §§ 669–693).

Peirce, C. S. (1893). A theory of probable inference. In *Peirce, C. S. 1931–1958* (Vol. 2, §§ 694–753).

Peters, B. G., & Nagel, M. (2022). *Zombie ideas: Why failed policy ideas persist.* https://www.ippapublicpolicy.org.

Piketty, T. (2014) [2013]. *Capital in the twenty-first century.* The Belknap Press of Harvard University Press.

Piketty, T. (2020). *Capital and ideology.* Harvard University Press.

Plato. (1990) [ca. 380–375BC]. *Plato's parmenides* (ed. by C.C. Meinwald). Oxford University Press.

Plato. (2009) [ca. 380–375BC]. *Phaedo* (ed. by D. Gallop), Oxford University Press.

Plato. (2010) [ca. 380–375BC]. *The Republic of Plato* (ed. by J. Adam) (Vol. 1). Cambridge University Press.

Plato. (2011) [ca. 380–375BC]. *The Republic of Plato* (ed. by J. Adam) (Vol. 2). Cambridge University Press.

Polanyi, K. (1957). *The great transformation.* Beacon.

Popper, K. R. (1959). *The logic of scientific discovery.* Harper Torchbooks.

Popper, K. R. (2007). *Conjectures and refutations.* Routledge.

Popper, K. R. (2009). *The two fundamental problems of the theory of knowledge.* Routledge.

Powell, J. L. (2015). Foucault power and culture. *International Journal of Humanities and Cultural Studies, 1*(4), 401–419.

Powell, J. L. (2020). *Theories of aging: New social horizons.* Nova Science.

Quiggin, J. (2012). *Zombie economics: How dead ideas still walk among us.* Princeton.

Rabinow, P. (Ed.) (2010). *Michel Foucault: The Foucault Reader.* Vintage.

Raphael, S., & Stoll, M. A. (2013). *Why are so many Americans in prison?* Russell Sage Foundation.

Remington, T. F. (2023). Socio-economic and geographic inequality of the effects of the pandemic. In: C. Aspalter (Ed.), *The covid-19 pandemic: Problems arising in health and social policy.* Springer.

Sawyer, W., & Wagner, P. (2023). *Mass incarceration: The whole pie, prison policy initiative.* https://www.prisonpolicy.org/reports/pie2023.html.

Schmitt, I., Nürnberger, A., & Lehrack, S. (2009). On the relation between fuzzy and quantum logic. In R. Seising (Ed.), *Views on fuzzy sets and systems from different perspectives.* Springer.

Schrödinger, E. (1935a). Die gegenwärtige Situation in der Quantenmechanik. *Naturwissenschaften, 23*(48 and 49): 807–812, 823–828, 844–849.

Schrödinger, E. (1935b). Discussion of probability relations between separated systems. *Proceedings of the Cambridge Philosophical Society, 31,* 555–563.

Schrödinger, E. (1936). Discussion of probability relations between separated systems II. *Proceedings of the Cambridge Philosophical Society, 32,* 446–451.

Schurz, G. (2008). Patterns of abduction. *Synthese, 164,* 201–234.

Seewald, B. (2020). *Rauschende Feste und Orgien bereiteten der Zivilisation den Weg*. https://www.welt.de/geschichte/article223294138/Goebekli-Tepe-Feste-und-Orgien-bereiteten-der-Zivilisation-den-Weg.html.

SF, Studyflix. (2023). *Heisenbergsche Unschärferelation*. https://studyflix.de/ingenieurwissenschaften/heisenbergsche-unscharferelation-2023.

Shields, H., Simpson, S., & Banks, L. (2019). *Einstein, Jung and the theory of everything*. Library X.

Smith, A. (1759). *The theory of moral sentiments*. A. Millar.

Smith, A. (1937) [1776]. *An inquiry into the nature and causes of the wealth of nations*. The Modern Library.

Smith, V. L. (2017). *The evidence of things not seen: Reflections on faith, science, and economics*. Acton Institute.

Smith, V. L., & Wilson, B. J. (2019). *Humanomics: Moral sentiments and the wealth of nations for the twenty-first century*. Cambridge University Press.

Surov, I. A., et al. (2021). Quantum semantics of text perception. *Scientific Reports, 11*, 4193. https://doi.org/10.1038/s41598-021-83490-9

Sutton, M. (2022). *Science fraud: Darwin's plagiarism of Patrick Matthew's theory*. Curtis Press.

Sutton, M., & Griffiths, M. (2023). The 'Patrick Matthew Effect' in science. In G. J. Curtis (Ed.), *Academic integrity in the social sciences*. Springer.

Swami Prabhupada, A. C. B. (1968). *Bhagavad Gita as it is*. Macmillan.

Swingle, B. (2018). Spacetime from entanglement. *Annual Review of Condensed Matter Physics, 9*, 345–435.

Tamvada, J. P., & Chowdhury, R. (2022). The irrationality of rationality in market economics: A paradox of incentives perspective. *Business & Society, 62*(3). https://doi.org/10.1177/00076503221101888.

Thompson, H. A. (2010). Why mass incarceration matters: rethinking crisis, decline, and transformation in Postwar American history. *The Journal of American History, 97*(3), 703–734.

Tilly, C. (2001). Relational origins of inequality. *Anthropological Theory, 1*(3), 355–372.

Van Kersbergen, K., & Vis, B. (2015). Three worlds' typology: Moving beyond normal science? *Journal of European Social Policy, 25*(1), 111–123.

Vredenburgh, D., & Brender, Y. (1998). The hierarchical abuse of power in work organizations. *Journal of Business Ethics, 17*, 1337–1347.

Vygotsky, L. S. (1962). *Thought and language*. MIT Press.

Vygotsky, L. S. (1978). *Mind in society: Development of higher psychological processes*. Harvard University Press.

Vygotsky, L. S. (1987). Thinking and speech. In R.W. Rieber, & A. S. Carton (Eds.), *The collected works of L.S. Vygotsky: Problems of general psychology* (Vol. 1). Plenum Press.

Wallace, A. R. (1855). On the law which has regulated the introduction of new species. *The Annals and Magazine of Natural History, 16*(2), 184–196.

Weber, M. (1952). *The protestant ethic and the spirit of capitalism*. Scribner.

Weber, M. (2012). The "Objectivity" of knowledge in social science and social policy. In H. H. Bruun, & S. Whimster (Eds.), *Max Weber*. Routledge.

Weber, M. (2019). Open and closed relationships. In D. B. Grusky (Ed.), *Social stratification*. Routledge.

Wendt, A. (2015). *Quantum mind and social science: Unifying physical and social ontology*. Cambridge University Press.

William, J. (1970). *The meaning of truth*. University of Michigan Press.

William, J. (1981). *Pragmatism* (ed. B. Kuklick), Hackett.

Wilson, R. A. (2016). *Quantum psychology: How brain software programs you and your world*. Hilaritas.

Winnicott, D. W. (1992). *Babies and their mothers*. Da Capo Lifelong Books.

Winnicott, D. W. (1994). *Talking to parents*. Da Capo Lifelong Books.

Worrall, J. (2002). Normal science and dogmatism, paradigms and progress: Kuhn 'versus' Popper and Lakatos. In T. Nickles (Ed.), *Thomas Kuhn.* Cambridge University Press.

Wu, B. (2023). *Website.* http://wubai.com.

Wu, S., Li, J., Zhang, P., & Zhang, Y. (2021). Natural language processing meets quantum physics: A survey and categorization. In *Proceedings of the 2021 conference on empirical methods in natural language processing*, (pp. 3172–3182), November 7–11. https://aclanthology.org/2021.emnlp-main.254.pdf.

Zeng, K., & Le Tendre, G. (1998). Adolescent suicide and academic competition in East Asia. *Comparative Education Review, 42*(4), 513–528.

Zimmermann, J. A. (2018). *Understanding the Heisenberg uncertainty principle.* https://www.thoughtco.com/the-heisenberg-uncertainty-principle-2699357.

Zohar, D., & Marshall, I. (1993). *The quantum society: Mind, physics and a new social vision.* Bloomsbury.